CEDAR MILL & BETHANY
LIBRARIES
D0455162

suicidal

suicidal

WHY WE KILL OURSELVES

jesse bering

THE UNIVERSITY OF CHICAGO PRESS
CHICAGO

The University of Chicago Press, Chicago 60637

Published 2018
Printed in the United States of America

27 26 25 24 23 22 21 20 19 18 1 2 3 4 5

ISBN-13: 978-0-226-46332-2 (cloth)
ISBN-13: 978-0-226-46346-9 (e-book)
DOI: https://doi.org/10.7208/chicago/9780226463469.001.0001

Library of Congress Cataloging-in-Publication Data
Names: Bering, Jesse, author.
Title: Suicidal: why we kill ourselves / Jesse Bering.
Description: Chicago: The University of Chicago Press, 2018. | Includes bibliographical references and index.
Identifiers: LCCN 2018021904 | ISBN 9780226463322 (cloth: alk. paper) | ISBN 9780226463469 (e-book)
Subjects: LCSH: Suicide.
Classification: LCC HV6545.B425 2018 | DDC 362.2/81—dc23
LC record available at https://lccn.loc.gov/2018021904

♾ This paper meets the requirements of ANSI/NISO Z39.48-1992 (Permanence of Paper).

FOR THE SUICIDAL PERSON IN ALL OF US

And so far forth death's terror doth affright,
He makes away himself, and hates the light
To make an end of fear and grief of heart,
He voluntarily dies to ease his smart.

Robert Burton, *The Anatomy of Melancholy* (1621)

Given the sensitive nature of the material in this book, I have not used any real names (unless otherwise stated), and I have changed physical descriptions, locations, and other features to ensure that no one is identifiable and their story is protected. This is because this is not a book about the individuals I have described, but about what we can learn from them and how they shape our lives.

contents

1

the call to oblivion

Just as life had been strange a few minutes before, so death was now as strange. The moth having righted himself now lay most decently and uncomplainingly composed. O yes, he seemed to say, death is stronger than I am.

Virginia Woolf, "The Death of the Moth" (1942)

Just behind my former home in upstate New York, in a small, dense pocket of woods, stood an imposing lichen-covered oak tree built by a century of sun and dampness and frost, its hardened veins criss crossing on the forest floor. It was just one of many such specimens in this copse of dappled shadows, birds, and well-worn deer tracks, but this particular tree held out a single giant limb crooked as an elbow, a branch so deliberately poised that whenever I'd stroll past it while out with the dogs on our morning walks, it beckoned me.

It was the perfect place, I thought, to hang myself.

I'd had fleeting suicidal feelings since my late teenage years. But now I was being haunted day and night by what was, in fact, a not altogether displeasing image of my corpse spinning ever so slowly from a rope tied around this creaking, pain-relieving branch. It's an absurd thought—that I could have observed my own dead body as if I'd casually stumbled upon it. And what good would my death serve if it meant having to view it through the eyes of the very same head that I so desperately wanted to escape from in the first place?

Nonetheless, I couldn't help but fixate on this hypothetical scene of the lifeless, pirouetting dummy, this discarded sad sack whose long-suffering owner had been liberated from a world in which he didn't truly belong.

Globally, a million people a year kill themselves, and many times that number try to do so. That's probably a hugely conservative estimate, too; for reasons such as stigma and prohibitive insurance claims, suicides and attempts are notoriously underreported when it comes to the official statistics. Roughly, though, these figures translate to the fact that someone takes their own life every forty seconds. Between now and the time you finish reading the next paragraph, someone, somewhere, will decide that death is a more welcoming prospect than breathing another breath in this world and will permanently remove themselves from the population.

The specific issues leading any given person to become suicidal are as different, of course, as their DNA—involving chains of events that one expert calls "dizzying in their variety"—but that doesn't mean there aren't common currents pushing one toward this fatal act. We're going to get a handle on those elusive themes in this book and, ultimately, begin to make sense of what remains one of the greatest riddles of all time: Why would an otherwise healthy person, someone even in the prime of their life, "go against nature" by hastening their death? After all, on the surface, suicide wouldn't appear to be a very smart Darwinian tactic, given that being alive would seem to be the first order of business when it comes to survival of the fittest.

But like most scientific questions, it turns out it's a little more complicated than that.

We won't be dealing here with "doctor-assisted suicide" or medical euthanasia, what Derek Humphrey in *Final Exit* regarded as "not suicide [but] self-deliverance . . . thoughtful, accelerated death to avoid further suffering from a physical disease." I consider such merciful instances of death almost always to be ethical and humane. Instead, we'll be focusing in the present book on those self-killings precipitated by fleeting or ongoing mental distress, namely, those that aren't the obvious result of physical pain or infirmity. Our primary analysis will center on the suicides of otherwise normal folks battling periodic depression or who suddenly find themselves in unexpected and overwhelming social circumstances. Plenty of suicides are linked to major psychiatric conditions (in which the person has a tenuous grasp of reality, such as in schizophrenia), but plenty aren't. And it's that everyday person dealing with suicidal thoughts—the suicidal person in all of us—who is the main subject of this book.

Benjamin Franklin famously quipped that "nine men in ten are would-be suicides." Maybe so, but some of us will lapse into this state more readily. It's now believed that around 43 percent of the variability in suicidal behavior among the general population can be explained by genetics, while the remaining 57 percent is attributable to environmental factors. When people who have a genetic predisposition for suicidality find themselves assaulted by a barrage of challenging life events, they are particularly vulnerable.

The catchall mental illness explanation only takes us so far. The vast majority of those who die by suicide, with some estimates as high as 90 percent,* have underlying psychiatric conditions, especially mood disorders such as depressive illness and bipolar disorder. (I have frequently battled the former, coupled with social anxiety.) But it's also true that not everyone with depression is suicidal, nor, believe it or not, is everyone who commits suicide depressed. According to one estimate, around 5 percent of depressed people will die by suicide, but about half a percent of the non-depressed population will end up taking their own lives too.

As for my own recurring compulsion to end my life, which flares up like a sore tooth at the whims of bad fortune, subsides for a while, yet always threatens to throb again, the types of problems that trigger these dangerous desires change over time. Edwin Shneidman, the famous suicidologist—yes, that's an actual occupation—had an apt term for this acute, intolerable feeling that makes people want to die: "psychache," he called it. It's like what Winona Ryder's character in the film *Girl, Interrupted* said after throwing back a fistful of aspirin in

*This oft-cited figure is highly contested, however, because it's derived primarily from postmortem analyses ("psychological autopsies"), which are almost certainly subject to hindsight bias. When mental health workers are given edited case histories of the same suicide victims *without knowing they've taken their own lives,* they are far less likely to see a mental illness.

a botched suicide attempt—she just wanted "to make the shit stop." And like a toothache, which can be set off by any number of packaged treats at our fingertips, psychache can be caused by an almost unlimited number of things in our modern world.

What made me suicidal as a teenager—the ever-looming prospect of being outed as gay in an intolerant small midwestern town—isn't what pushes those despairing buttons in me now. I've been out of the closet for twenty years and with my partner, Juan, for over a decade. I do sometimes still wince at the memory of my adolescent fear regarding my sexual orientation, but the constant worry and anxiety about being forced prematurely out of the closet are gone now.

Still, other seemingly unsolvable problems continue to crop up as a matter of course.

*

What drew me to those woods behind my house not so long ago was my unemployment. I was sorely unprepared for it. Not long before, I'd enjoyed a fairly high status in the academic world. Frankly, I was spoiled. And lucky. That part I didn't realize until much later. I'd gotten my first faculty position at the University of Arkansas straight out of grad school. Then, at the age of thirty, I moved to Northern Ireland, where I ran my own research center for several years at the Queen's University Belfast.

Somewhere along the way, though, my scholarly ambitions began to wear thin.

It was a classic case of career burnout. By the time I was thirty-five, I'd already done most of what I'd set out to do: I was publishing in the best journals, speaking at conferences all over the world, scoring big grants, and writing about my research (in religion and psychology) for popular outlets. If I were smart, I'd have kept my nose to the grindstone. Instead, I grew restless. "Now what?" I asked myself.

The prospect of doing slight iterations of the same studies over and over became a nightmare, the academic's equivalent of being stuck in a never-ending time loop. Besides, although controversial issues like religion are never definitively settled, I'd already answered my main research question, at least to my own satisfaction. (Question: "What are the odds that religious ideas are a product of the human mind?" Answer: "Pretty darn high.")

With my professorial aspirations languishing, I began devoting more and more time to writing popular science essays for outfits such as *Scientific American, Slate, Playboy,* and a few others. My shtick was covering the salacious science beat. If you'd ever wondered about the relationship between gorilla fur, crab lice, and human pubic hair, about the mysterious psychopharmacological properties of semen, or why our species' peculiar penis is shaped like it is, I was your man. In fact, I wrote that very book: *Why Is the Penis Shaped Like That?*

The next book I was to write had an even more squirm-inducing title: *Perv: The Sexual Deviant in All of Us*. Ever wonder why amputees turn on some folks, others can't keep from having an orgasm when an attractive passerby lapses into a sneezing fit, or why women are generally kinkier than men? Again, I was your clickable go-to source.

Now, perhaps I should have thought more about how, in a conservative and unforgiving academic world, such subject matter would link my name inexorably with unspeakable things. Sure, my articles got page clicks. My books made you blush at Barnes & Noble. But these titles aren't exactly ones that university deans and provosts like to boast about to donors. Once you go public with the story of how you masturbated as a teenager to a wax statue of an anatomically correct Neanderthal (I swear it made sense in context), there is no going back. You can pretty much forget about ever getting inducted into the Royal Society. "Oh good riddance," I thought. Being finally free to write in a manner that suited me—and with my very own soapbox to say the things I'd long wanted to say about society's soul-crushing hypocrisy—was incredibly appealing.

There was also the money. I wasn't getting rich, but I'd earned large enough advances with my book deals to quit my academic job, book a one-way ticket from Belfast back to the U.S., and put a deposit down on an idyllic little cottage next to a babbling brook just outside of Ithaca. Back then, the dark patch of forest behind the house didn't seem so sinister; it was just a great place to walk our two border terriers, Gulliver and Uma, our rambunctious Irish imports. The whole domestic setting seemed the perfect little place to build the perfect little writing life—a fairy tale built on the foundations of other people's "deviant" sexualities.

You can probably see where this is heading. Juan, the more practical of us, raised his eyebrows early on over such an impulsive and drastic career move. By that I mean he was resolutely set against it. "What are you going to do after you finish the book?" he'd ask, sensing doom on the horizon.

"Write another book I guess. Maybe do freelance. I can always go back to teaching, right? C'mon, don't be such a pessimist!"

"I don't know," Juan would say worriedly. But he also realized how unhappy I was in Northern Ireland, so he went along, grudgingly, with my loosely laid plans.

✷

I wouldn't say my fall from grace was spectacular. But it was close. If nothing else, it was deeply embarrassing. It's hard to talk about even now that I'm, literally, out of the woods.

That's the thing. Much of what makes people suicidal *is* hard to talk about. Shame plays a major role. Even suicide notes, as we'll learn, don't always key us in to the real reason someone opts out of existence. (Forgive the glib euphemisms; there are only so many times one can write the word "suicide" without expecting readers' eyes to glaze over.) If I'll be asking others in this book to be honest about their feelings, though, it would be unfair for me to hide the reasons for my own self-loathing and sense of irredeemable failure during this dark period.

It's often at our very lowest that we cling most desperately to our points of pride, as though we're trying to convince not only others, but also ourselves, that we still have value.

Once, long ago, when I was about twenty, I met an old man of about ninety who carried around with him an ancient yellowed letter everywhere he went. People called him "the Judge."

"I want to show you something, young man," he said to me after a dinner party, reaching a shaky hand into his vest pocket to retrieve the letter. "See that?" he asked, beaming. A twisted arthritic finger was pointing to a typewritten line from the Prohibition era. As I tried to make sense of the words on the page, he studied my gaze under his watery pink lids to be sure it was really sinking in. "It's a commendation from Franklin D. Roosevelt, the governor of New York back then. Says here, see, says right here I was the youngest Supreme Court Justice in the state. Twenty. Eight. Years. Old." With each punctuated word, he gave the paper a firm tap. "Whaddaya think of that?"

"That's incredibly impressive," I said.

And it was. In fact, I remember being envious of him. Not because of his accomplished legal career, but because, as I so often have been in my life, I was suicidal at the time; and unlike me, he hadn't long to go before slipping gently off into that good night.

One of the cruelest tricks played on the genuinely suicidal mind is that time slows to a crawl. When each new dawn welcomes what feels like an eternity of mental anguish, the yawning expanse between youth and old age might as well be interminable Hell itself.

But the point is that when we're thrown against our wishes into a liminal state—that reluctant space between activity and senescence, employed and unemployed, married and single, closeted and out, citizen and prisoner, wife and widow, healthy person and patient, wealthy and broke, celebrity and has-been, and so on—it's natural to take refuge in the glorified past of our previous selves. And to try to remind others of this eclipsed identity as well.

Alas, it's a lost cause. Deep down, we know there's no going back. Our identities have changed permanently in the minds of others. In the real world (the one whose axis doesn't turn on cheap clichés and self-help canons about other people's opinions of us not mattering), we're inextricably woven into the fabric of society.

For better or worse, our well-being is hugely dependent on what others think we are.

Social psychologist Roy Baumeister, whom we'll meet again later on, argues that idealistic life conditions actually heighten suicide risk because they create unreasonable standards for personal happiness. When things get a bit messy, people who have led mostly privileged lives—those seen by society as having it made—have a harder time coping with failures. "A reverse of fortune, as society is constituted," wrote the eighteenth-century thinker Madame de Staël, "produces a most acute unhappiness, which multiplies itself in a thousand different ways. The most cruel of all, however, is the loss of the rank we occupied in the world. Imagination has as much to do with the past, as with the future, and we form with our possessions an alliance, whose rupture is most grievous."

Like the Judge, I was dangerously proud of my earlier status. The precipitous drop between my past and my present job footing was discombobulating. I wouldn't have admitted it then, or even known I was guilty of such a cognitive crime, but I also harbored an unspoken sense of entitlement. Now, I felt like Jean-Baptiste Clamence in *The Fall* by Albert Camus. In the face of a series of unsettling events, the successful Parisian defense attorney watches as his career, and his entire sense of meaning, goes up in smoke. Only when sifting through the ashes are his biases made clear. "As a result of being showered with blessings," Clamence observes of his worldview till then,

> I felt, I hesitate to admit, marked out. Personally marked out, among all, for that long uninterrupted success. I refused to attribute that success to my own merits and could not believe that the conjunction in a single person of such different and such extreme virtues was the result of chance alone. This is why in my happy life I felt somehow that that happiness was authorized by some higher decree. When I add that I had no religion you can see even better how extraordinary that conviction was.

Similarly, what I had long failed to fully appreciate were the many subtle and incalculable forces behind my earlier success, forces that had always been beyond my control. I felt somehow, what is the word . . . charmed is too strong, more like fatalistic. The reality was that I was like everyone else, simply held upright by the brittle bones of chance. And now, they threatened to give way. I'd worked hard, sure, but again, I'd been lucky. Back when I'd earned my doctoral degree, the economy wasn't so gloomy and there were actually opportunities. I was also doing research on a hot new topic—my PhD dissertation was on children's reasoning about the afterlife—and I was eager to make a name for myself in a burgeoning field. Now, eleven years later, having turned my back on the academy, fresh out of book ideas, along with a name pretty much synonymous with penises and pervs, it was a very different story. Career burnout? Please. That's a luxury for the employed.

I just needed a steady paycheck.

*

The rational part of my brain assured me that my present dilemma was not the end of the world. Still, the little that remained of my book advance was drying up quickly, and my freelance writing gigs, feverishly busy as they kept me, didn't pay enough to live on. Juan, who'd been earning his master's degree in library science, was forced to take on a minimum-wage cashier job at the grocery store. He never said "I told you so." He didn't have to.

I knew going in that the grass wouldn't necessarily be greener on the other side of a staid career, but never did I think it could be scorched earth. That perfect little cottage? It came with a mortgage. We didn't have kids, but we did have two bright-eyed terriers and a cat named Tommy to feed and care for. Student loans. Taxes. Fuel. Credit cards. Electricity. Did I mention I was an uninsured Type I diabetic on an insulin pump? My blinkered pursuit of freedom to write at any cost was starting to have potentially fatal consequences.

Doing what you love for a living is great. But you know what's even more fun? Food.

The irrational part of my brain couldn't see how this state of affairs, which I'd stupidly, selfishly put us into, could possibly turn out well. Things were only going to get worse. Cue visions of foreclosure, confused, sad-faced, whimpering pets torn asunder and kenneled (or worse), loving family members, stretched to the limit already themselves, arguing with each other behind closed doors over how to handle the "situation with Jesse." Everyone, including me, would be better off without me; I just needed to get the animals placed in a loving home and Juan to start a fresh, unimpeded life back in Santa Fe, where he'd been living when we first met.

"You're such a loser," I'd scold myself. "You had it made. Now look at you."

Asshole though this internal voice could be, it did make some good points. What if *that* was the rational part of my brain, I began to wonder, and the more optimistic side—the one telling me it was all going to be okay—was delusional? After all, in the fast-moving world of science, I was now a dinosaur. I hadn't taught or done research for years. I'd also burned a lot of bridges due to my, er, penchant for sensationalism. An air of Schadenfreude, which I'm sure I'd rightfully earned from some of my critics, would soon be palpable.

Overall, I felt like persona non grata among all the proper citizens surrounding me, all those deeply rooted trees that so obviously belonged to this world. Even the weeds had their place. But me? I didn't belong. I was, in point of fact, simultaneously over- and underqualified for everything I could think of, saddled with an obscure advanced degree and absolutely no practical skills. And of course I might as well be a registered sex offender with the titles of my books and articles (among the ones I was working on at the time, "The Masturbatory Habits of Priests" and "Erotic Vomiting"). I envied the mailman, the store clerk, the landscaper . . . anyone with a clear purpose.

*

Meanwhile, the stark contrast between my private and public life only exacerbated my despondency. From a distance, it would appear that my star was rising. I was giving talks at the Sydney Opera House, being interviewed regularly by NPR and the BBC, and getting profiled in the *Guardian* and the *New York Times*. Morgan Freeman featured my earlier work on religion for his show *Through the Wormhole*. Meanwhile, over in the UK, the British illusionist Derren Brown did the same on his televised specials. My blog at *Scientific American* was nominated for a Webby Award. Dan Savage, the famous sex advice columnist, tapped me to be his substitute columnist when he went away on vacation for a week. I even did the late-night talk show circuit. Chelsea Handler brazenly asked me, on national television, if

I'd have anal sex with her. (I said yes, by the way, but I was just being polite.) A big Hollywood producer acquired the film option rights to one of my *Slate* articles.

With such exciting things happening in my life, how could I possibly complain, let alone be suicidal? After all, most writers would kill (no pun intended) to attract the sort of publicity I was getting. "Oh, boo-hoo," I told myself. "You've sure got it rough. Let's ask one of those new Syrian refugees how they feel about your dire straits, shall we? How about that nice old woman up the road vomiting her guts out from chemo?" A close friend from my childhood had just had a stroke and was posting inspirational status updates on his Twitter account as he learned how to walk again, #trulyblessed. What right did I have to be so unhappy?

This kind of internal self-flagellation, like reading a never-ending scroll of excoriating social media comments projected onto my mind's eye, only made being me more insufferable. I ambled along for months this way, miserable, smiling like an idiot and popping Prozac, hoping the constant gray drizzle in my brain would lift before the dam finally flooded and I got washed up into the trees behind the house.

No one knew it. At least, not the full extent of it.

From the outside looking in, even to the few close friends I had, things were going swimmingly. "When are you going to be on TV again?" they'd ask. "Where to next on your book tour?" Or "Hey, um, interesting article on the history of autofellatio."

All was illusion. The truth is these experiences offered little in the way of remuneration. The press didn't pay. The public speaking didn't amount to much. And the film still hasn't been made.

My outward successes only made me feel like an impostor. Less than a week after I appeared as a guest on *Conan,* I was racking my head trying to think of someone, anyone, who could get me a gun to blow it off. Yet look hard as you might at a recording of that interview from October 16, 2013, and you won't see a trace of my crippling worry and despair. What does a suicidal person look like? Me, in that *Conan* interview.

✷

Here's the trouble. We're not all ragingly mad, violently unstable, or even obviously depressed. Sometimes, a suicide seems like it comes out of nowhere. But that's only because so many of us would rather go to our graves keeping up appearances than reveal we're secretly coming undone.

In response to an article in *Scientific American* in which I'd shared my personal experiences as a suicidal gay teenager (while keeping my current mental health issues carefully under wraps), one woman wrote to me about the torturous divide between her own public persona and private inner life. "It's difficult to admit that at age 34," she explained,

> with a young daughter, a graduate degree in history, divorced, and remarried to my high school love, that I'm Googling suicide. But what the world doesn't see is years of fertility issues, childhood rape, post-traumatic stress disorder, a failing marriage, a custody battle, nonexistent career, mounds of debt, and a general hatred of myself. Depression is a secret tomb that no one sees but you . . . being dead but yet alive.

She's far from alone. There are more people walking around this way, "dead but yet alive," than anyone realizes.

In my case, being open about my persistent suicidal thoughts at a time when readers' perception of me as a good, clearheaded thinker meant the difference between a respectable middle age and moving into my elderly father's basement and living off cans of SpaghettiOs. It just wasn't something I was willing to do at the time. Who'd buy a book by an author with a mood disorder, a has-been academic, and a self-confessed sensationalist who can't stop thinking about killing himself, and take him seriously as an authoritative voice of reason?

I don't blame anyone for missing the signs. What signs? Anyway, regrettably, I've done the same. The man who'd designed my website, a sweet, introverted IT guy also struggling to find a job, overdosed while lying on his couch around this time. His landlord found him three days later with his two cats standing on his chest, meowing. I was unnerved to realize that despite our mutual email pleasantries, we'd both in fact wanted to die.

We're more intuitive than we give ourselves credit for, but people aren't mind readers. We come to trust appearances; we forget that others are self-contained universes just like us, and the deep rifts forming at the edges go unnoticed, until another unreachable cosmos "suddenly" collapses. In the semiautobiographical *The Book of Disquiet,* Fernando Pessoa describes being surprised upon learning that a young shop assistant at the tobacco store had killed himself. "Poor lad," writes Pessoa, "so he existed too!"

> We had all forgotten that, all of us; we who knew him only about as well as those who didn't know him at all. . . . But what is certain is that he had a soul, enough soul to kill himself. Passions? Worries? Of course. But for me, and for the rest of humanity, all that remains is the memory of a foolish smile above a grubby woollen jacket that didn't fit properly at the shoulders. That is all that remains to me of someone who felt deeply enough to kill himself, because, after all[,] there's no other reason to kill oneself.

These dark feelings are inherently social in nature. In the vast majority of cases, people kill themselves because of other people. Social problems—especially, a hypervigilant concern with what others think or will think of us if only they knew what we perceive to be some unpalatable truth—stoke a deadly fire.*

✷

*In one telling old study using a projective test that involved presenting participants with an array of background images (a bridge, a living room, a street, and so on) along with a wide range of cutouts of miscellaneous human and animal figures, and then asking them to use only a limited subset of these pictures to tell a story, suicidal participants incorporated significantly more human characters into their fictional stories than did schizophrenic and control participants, suggesting an increased concern with other people.

Fortunately, suicide isn't inevitable. As for me, it's funny how things turned out. (And I mean "funny" in the way a lunatic giggles into his hand, because this entire wayward career experience must have knocked about five years off my life.) Just as things looked most grim, I was offered a job in one of the most beautiful places on the planet: the verdant wild bottom of the South Island in New Zealand. In July 2014 Juan, Gulliver, Uma, Tommy, and I—the whole hairy, harried family—packed up all of our earthly possessions, drove across country in a rented van, and flew from Los Angeles to Dunedin, where I'd been hired as the writing coordinator in a new Science Communication department at the University of Otago.

Ironically, I wouldn't have been much of a candidate had I not devoted a few solid nail-biting years to freelancing. I'll never disentangle myself from my reputation as a purveyor of pervy knowledge, but the Kiwis took my frank approach to sex with good humor.

Outside our small home on the Otago Peninsula, I'm serenaded by tuis and bellbirds; just up the road, penguins waddle from the shores of an endless ocean each dusk to nest in cliff-side dens, octopuses bobble at the harbor's edge, while dolphins frolic and giant albatrosses the size of small aircraft soar overhead. At night the Milky Way is so dense and bright against the inky black sky, I can almost reach up and stir it, and every once in a while, the aurora australis, otherwise known as the southern lights, puts on a spectacular multicolored display. The dogs are thriving. The cat is purring. Juan has a great new job.

I therefore whisper this to you as though the cortical gods might conspire against me still: I'm currently "happy" with life.

I use that word—*happy*—with trepidation. It defines not a permanent state of being but slippery moments of non-worry. All we can do, really, is try to maximize the occurrence of such anxiety-free moments throughout the course of our lives; a worrisome mind is a place where suicide's natural breeding ground, depression, spreads like black mold.

Personally, I'm all too conscious of the fact that had things gone this way or that but by a hairbreadth, my own story might just as well have ended years ago at the end of a rope on a tree that grows

8,000 miles away. Whether I'd have gone through with it is hard to say. I don't enjoy pain, but I certainly wanted to die, and there's a tipping point where the agony of living becomes worse than the pain of dying. It would be naive of me to assume that just because I called the universe's bluff back then, my suicidal feelings have been banished for good. As I write this, I'm forty-two years of age, and so there's likely plenty of time for those dark impulses to return. Perhaps they're merely lying in wait for the next unmitigated crisis and will come back with a vengeance. Also, according to some of the science we'll be examining, I possess almost a full complement of traits that make certain types of people more prone to suicide than others. Impulsive. *Check.* Perfectionist. *Check.* Sensitive. Shame-prone. Mood-disordered. Sexual minority. Self-blaming. *Check. Check. Check. Check. Check.*

We're used to safeguarding ourselves against external threats and preparing for unexpected emergencies. We diligently strap on our seat belts every time we get in a car. We lock our doors before bed. Some of us even carry weapons in case we're attacked by a stranger. Ironic, then, that statistically we're far more likely to perish intentionally by our own hand than to die of causes that are more obviously outside of our control. In fact, historically, suicide has accounted for more deaths than all wars and homicides combined.

When I get suicidal again—not if, but when—I want to be armed with an up-to-date scientific understanding that allows me to critically analyze my own doomsday thoughts or, at the very least, to be an informed consumer of my own oblivion. I want you to have that same advantage. That's largely why I have written this book . . . to reveal the psychological secrets of suicide, the tricks our minds play on us when we're easy emotional prey. It's also about leaving our own preconceptions aside and instead considering the many different experiences of those who've found themselves affected somehow—whether that means getting into the headspaces of people who killed themselves or are actively suicidal, those bereaved by the suicide death of a loved one, researchers who must quarantine their own emotions to study suicide objectively, or those on the grueling front lines of prevention campaigns. Finally, we'll be exploring some challenging, but fundamental, questions about how we wrestle with the ethical questions surrounding suicide, and how our intellect is often at odds with our emotions when it comes to weighing the "rationality" of other people's fatal decisions.

✷

Unlike most books on the subject, this one doesn't necessarily aim to prevent all suicides. My own position, for lack of a better word, is nuanced. In fact, I tend to agree with the Austrian scholar Josef Popper-Lynkeus, who remarked in his book *The Right to Live and the Duty to Die* (1878) that, for him, "the knowledge of always being free to determine when or whether to give up one's life inspires me with the feeling of a new power and gives me a composure comparable to the consciousness of the soldier on the battlefield."

The trouble is, being emotionally fraught with despair can also distort human decision making in ways that undermine a person's ability to decide intelligently "when or whether" to act. Because despite our firm conviction that there's absolutely no escape from that seemingly unsolvable, hopeless situation we may currently find ourselves in, we're often—as I was—dead wrong in retrospect. "Never kill yourself while you are suicidal" was one of Shneidman's favorite maxims.

Intellectualizing a personal problem is a well-known defense mechanism, and it's basically what I'll be doing in this book. Some might see this coldly scientific approach as a sort of evasion tactic for avoiding unpleasant emotions. Yet with suicide, I'm convinced that understanding suicidal urges, from a scientific perspective, can keep many people alive, at least in the short term. My hope is that knowing how it all works will help us to short-circuit the powerful impetus to die when things look calamitous. I want people to be able to recognize when they're under suicide's hypnotic spell and to wait it out long enough for that spell to wear off. Acute episodes of suicidal ideation rarely last longer than twenty-four hours. Education may not always lead to prevention, but it certainly makes for good preparation. And for those of you trying to understand how someone you loved or cared about could have done such an inexplicable thing as to take their own life, my hope is that you'll benefit, too, from this examination of the self-destructive mind and how we, as a society, think about suicide.

There's only so much ground that can be covered in a book of this sort, of course. Well over a century's worth of scientific theory and research on the subject of suicide makes this a formidable topic to explore. Because my own research area is in the field of social cognition, the academic literature I'll be examining is mostly from that broad discipline. Admittedly, I'll be cherry-picking. But I make no apologies for that. The particular theorists and studies that I've chosen to highlight have helped me to see the issues more clearly and have even offered me some therapeutic insight. These are the ideas that, in my view, are worth knowing.

Needless to say—but I'll say it anyway—what lies ahead won't always be easy reading. Unfortunately, it comes with the territory. Still, please, brace yourself. If you're coming into this fresh in the wake of a tragedy, or things are simply a little too raw right now, consider coming back to the book later. It's here when you're ready to read it.

For the rest of you settling in for the ride, meanwhile, let's begin with a basic, but deceptively significant, question: Is suicide uniquely human? Camus, for his part, saw suicide as "the one truly serious philosophical problem," regarding it as something that sets humans apart from all other species. If he was right, then what could it be, precisely, that saves other animals from becoming their own deliberate executioners? Let's take a closer look.

2

unlike the scorpion girt by fire

For all cats have this particularity, each and every one, from the meanest alley sneaker to the proudest, whitest she that ever graced a pontiff's pillow—we have our smiles, as it were, painted on. Those small, cool, quiet Mona Lisa smiles that smile we must, no matter whether it's been fun or it's been not.

Angela Carter, *Puss-in-Boots* (1979)

One day many years ago, somewhere in northwest Arkansas, a litter of kittens was born. For whatever reason—the whims of an overly affectionate child, perhaps, or more likely an impatient owner indifferent to the basics of feline husbandry—the animals were removed far too early from their mother. That much was as obvious as the razor-sharp nails that dug rhythmically, ecstatically, into my flesh the first time I held Tommy, a six-month-old tuxedo cat, in the visitors' lounge at the county animal shelter. Such "kneading" behavior, say cat experts, is a remnant of what young kittens do with their paws against the mother cat's teats to facilitate milk letdown while they're suckling. Pluck the little ones away while they're still in the nursing stage and the habit is fixed for life.

Seeing, at first, no more than a nondescript cat whose doppelgänger would be easy enough to find among all the similar black-and-whites in the world—several were in this very room—I quickly broke his gaze and moved on to a more fetching Abyssinian with a ringed powdery coat resembling that of a wild hare. Still, I couldn't help but feel this other (as much as I dislike this vaguely supernatural term) *presence* following my every move.

Actually, there was nothing supernatural about it; Tommy's nonstop meowing was . . . how should I describe it? It was the caterwaul of a constipated seagull, tuned precisely to unnerve the human eardrum and cause a person to do very irrational things, such as, for instance, open the door to an animal's cage in a desperate attempt to calm it down.

"Now *that's* a cat and a half," said my friend Patty, watching as the oversize tomcat melted into my arms, its howling switching instantaneously to raucous purring. "Oh, come on," Patty continued, "get him. Look at that. He *loves* you." There's no denying this cat and I had chemistry. If I let him get any closer, I was afraid he was going to latch on to my nipple.

"Fine," I said, secretly smitten. "I suppose I could use a good mouser."

The last thing on my mind was that this cat was suicidal. Yet the following day, I found myself in the middle of the impenetrable forest behind my rustic cabin in the Ozark National Forest, craning my neck to a canopy so high and unbroken it all but blotted out the sky. I'd been attempting to localize the precise source of a disembodied meow. I knew right away it was Tommy because it had the same hysterical ring I'd heard at the shelter.

When I finally spotted him, the cat was hunkered down on a bowing branch at the tippy top of a massive pine, so far up that I had to squint to be sure it was Tommy and not, say, a rabid bloated possum. "Now what?" I thought as we stood there staring at each other across half the length of a football field: him, down, me, up. The branch was bending worrisomely under his formidable weight. A warm, ominous wind was beginning to stir, and I watched as a few sleepy box turtles crawled up from the gully in search of higher ground, a sign of a gathering storm. I stood there scratching my head. "What am I supposed to do, call the fire department?" I thought. "Wait, do people actually do that? Is that what people do?"

I did it.

"Nine-one-one, what's your emergency?" The voice was all business.

"Um, yes, I don't know if this qualifies as an 'emergency' per se," I prevaricated, "but you see, my cat is stuck at the top of a tree and doesn't seem to be able to get back down." I covered the receiver. "Don't jump!" I screamed up at Tommy. "We're going to figure this out!"

After a pregnant pause—one in which I could almost hear neurons busily assembling synapses—the operator returned to the line. "Sir," she said in a way that made me picture her standing there straightening out her dress, having had just about enough of random callers' feline catastrophes. "This isn't Mayberry. You'll have to phone your vet. Or something."

At this point I should probably come clean. I've been winding you up. Not that this didn't happen. Every breath of it is true. But perhaps it's a stretch to assume that Tommy was suicidal, which is to say, intent on ending his life that afternoon. Then again, stranger things have happened than a cat dying from a fall. Who are we to say such deaths are accidental?

In any event, here's how the most recent edition of the *Diagnostic and Statistical Manual of Mental Disorders*, the *DSM-5*, defines "suicide":

> An act with a fatal outcome, deliberately initiated and performed by the [individual] with the knowledge or expectation of its fatal outcome.

Let's look at this objectively. If Tommy had plummeted from the tree, could we rule out the possibility that he'd climbed up there because he, say, was still feeling dejected about being abandoned by his former owner and had every intention of jumping to his death—until, that is, he looked down to see the worry on my face and realized we really were going to be best mates?*

*Actually, now that I think about it, my dog had always acted rather suspiciously too. A voracious eater, such was her nimble intelligence in this domain that, ever since

Well, you can't disprove such suppositions. But as critical thinkers trying to understand the natural foundations of suicide, are these kinds of humanlike attributions really the most plausible? After all, it's also possible—some would say probable—that Tommy was simply exploring his new environment and got in over his head. Also, here's some trivia: apparently young neutered male cats are the ones most likely to get themselves stuck in sky-high trees. And like Tommy, once they get it out of their system, they never do it again. It's a rite of passage that continues to confound scientists. Still honing their squirrel-hunting skills is my wager.

And for future reference, a thunderstorm and an open can of tuna placed at the base of the tree is all you need to get a bedraggled cat with a death wish to make its way down, backward, to terra firma.

*

The somewhat pedantic admonishment to always go with the least complicated of explanations for an animal's behavior is known as Morgan's canon. C. Lloyd Morgan was a British ethologist whose name, to this day, makes many scholars cringe because of what looks on the surface to be a smug attitude of human superiority. "In no case is an animal activity to be interpreted in terms of higher psychological processes," Morgan famously wrote in 1903 as a sort of addendum to Occam's razor, "if it can be fairly interpreted in terms of processes which stand lower in the scale of psychological evolution

---

returning home one day to find an empty quiche container and a tub of butter on the floor, I'd had to secure the refrigerator with electrical tape whenever I went out. Once, in a very scary incident, she'd jumped on the table to gorge on Halloween candy and brownies, and I'd been forced to deliver an emergency emetic down her throat. Here is an incomplete list of other items she'd scandalously ingested over the years: an enormous scented candle that smelled like sugar cookies, my small cousin's soiled diapers, several pounds of raw crawfish (along with half a container of Cajun seasoning), and a Kleenex containing a used condom that my brother had just discarded in the trash can following a surreptitious reunion with his then-fiancée at our mom's house. After a decade of fishing around in her mouth to extract such atrocities, odd how it never occurred to me that maybe she was trying to poison herself. I'm also beginning to see a depressing common denominator for my suicidal pets here: *me.*

and development." He'd be positively turning over in his grave over Tommy's "suicide" excursion.

Earlier in his career, in fact, Morgan had even weighed in on the question of whether other animals ever knowingly take their own lives. Back then the classic example of nonhuman suicide wasn't, say, lemmings or whales (more on those guys soon) but scorpions. As the science historians Edmund Ramsden and Duncan Wilson discuss in their fascinating look into the strangely politicized history surrounding animal suicide, this was largely the result of a popular poem written by the Romantic poet Lord Byron, one that had taken hold of the British imagination with its florid vision of a doomed, flame-trapped scorpion as an analogy for a person in hopeless circumstances. Composed originally in 1813, *The Giaour* portrays a creature so noble and rational that when surrounded by an inescapable ring of fire, it would sooner thrust its stinger into its own back and kill itself than be burned alive. Byron penned:

> The Mind, that broods o'er guilty woes,
> Is like the Scorpion girt by fire,
> In circle narrowing as it glows
> The flames around their captive close,
> Till inly search'd by thousand throes,
> And maddening in her ire,
> One sad and sole relief she knows,
> The sting she nourish'd for her foes,
> Whose venom never yet was vain,
> Gives but one pang, and cures all pain,
> And darts into her desperate brain.—
> So do the dark in soul expire,
> Or live like Scorpion girt by fire.

Nonsense. That was Morgan's cantankerous view on this romanticized portrayal of the ill-fated scorpion. Yes, it may sometimes appear that this is what's happening with a scorpion when you encircle it with flames, he pointed out, but on closer examination, the animal is really just striking at its own back in an effort to relieve the source of an "irritating" heat. In some rare instances, it might sting itself in the midst of all that flailing around, but not with conscious suicidal intent.

Morgan arrived at this conclusion by conducting what remains, as far as I can tell, the only study in the annals of scholarly peer-reviewed history in which a living organism was systematically goaded into killing itself . . . or, at least, given every opportunity to do so, provided it had the intellectual wherewithal. In his article "Suicide of Scorpions," appearing in the journal *Nature* in 1883, Morgan reported a series of sadistic laboratory tests, each designed to induce so much pain in these poor arachnids that if they did have the capacity to end their own lives, now would be the perfect time to do so. His torture techniques included such creative methods as "burning phosphorous on the creature's body," "placing in burning alcohol," "treating with a series of electric shocks," and, of course, "surrounding with fire or red hot embers" (which, contrary to Byron's poem, merely prompted Morgan's subjects to walk without hesitation through the fire to make their escape).

Countless scorpion lives were lost in this endeavor, but not a single one of them the result of suicide by self-stinging. "I think it will be admitted that some of these experiments were sufficiently barbarous," wrote Morgan, "to induce any scorpion who had the slightest suicidal tendency to find relief in self-destruction," adding parenthetically that the phosphorous test in particular was "positively sickening."

Perhaps a better starting point would have been first determining if scorpions are immune to their own venom, rendering the entire premise of scorpion suicide moot.

Turns out, they are.*

*

Few saw Morgan as a hero for his efforts. In fact, to a new class of progressive thinkers, many of whom were involved in the fledgling

*Although some experts still hold that if the venom is infused directly into the animal's basal ganglia, this can sometimes be fatal.

animal rights movement in Britain, Morgan was the ultimate killjoy, because unlike his contemporaries at the time, he had no patience for anecdotes and flighty notions of animals being conscious like us. Those people, by contrast, found inspiration in one of Morgan's nemeses, the evolutionary biologist George Romanes, whose charming "argument by analogy" seemed more humane than Morgan's canon.

"Starting from what I know of the operations of my own individual mind," Romanes boldly reasoned in 1882, "and the activities which in my own organism they prompt, I proceed by analogy to infer from the observable activities of other organisms what are the mental operations that underlie them." In so doing, Romanes was simply formalizing what we all do anyway: anthropomorphize. And, anyway, Romanes's theory also seemed more scientifically informed in the wake of Charles Darwin's theory of natural selection. After all, Darwin himself concluded in *The Descent of Man* (1871) that he saw "no fundamental difference" in the psychological abilities of humans and other animals.

So who was right, Morgan or Romanes?

Neither. And both.

It's scientifically naive to assume that humans have a monopoly on consciousness, but it's equally naive to think that other animals experience the world in the same way that we do. All organisms emerged from the same primordial ooze, and thus all neurophysiology is made of the same essential stuff. It was, indeed, all a kind of neural Pangaea eons ago, but some psychological landmasses, especially those 220 million years apart, no longer connect in any meaningful way. Consider that we parted ways with our closest living relatives—chimpanzees—five to seven million years ago. That's a long time, even on a geologic scale. We may be the only hominids remaining, but no less than twenty intermediate species of human have come and gone in that long interval.

Even a cursory glance between these two extant brains, that of chimp and modern human, will show that in terms of structure (and therefore function), much has happened since we last shared a common ancestor. Many of those differences are packaged in the frontal cortex, the brain hub where things such as perspective-taking and behavioral inhibition occur. Do these "human cognitive specializations," as the comparative psychologist Daniel Povinelli refers to this suite of unique social cognitive traits, make us special among all other creatures? Not in any value-laden religious sense, they don't. But it does make us biologically different, just like every species is unique in some way. And these differences have major implications for our understanding of suicide.

Darwin was right about mental continuity, and Romanes's argument by analogy is plausible in certain respects too, but presuming that there's "no fundamental difference" between your subjective reality and that of the housefly peering down at you through all of its three thousand eyes at this very moment doesn't sound very convincing.

This isn't a new argument. In 1974 a philosopher named Thomas Nagel wrote a now-classic essay titled "What Is It Like to Be a Bat?" Nagel's short answer: "How the heck should I know?" Paraphrased, of course. What Nagel was getting at was the fact that entering a bat's private mind—being a bat—requires an inner frame of reference that is generated by a singularly evolved bat neurocircuitry (such as that used in echolocation), which human brains simply don't possess in any clear way, shape, or form. Sure, we can attempt to approach the phenomenology of being a bat through the linguistic gymnastics of analogies and metaphors, but it's still only a simulation constructed by a very un-bat-like human brain.

That doesn't keep us from asking, and probably erroneously answering, questions about other species' impenetrable worldviews. But the truth is that it's impossible to know why the chicken crossed the road. Or why the cat climbed the tree, for that matter.

But let's not get stuck already in a philosophical quagmire. For now, the important point is that we can't help but put ourselves in the shoes—or hooves, paws, or pincers (in the case of scorpions)—of other organisms. It's how we're wired.

*

Many researchers believe that what makes us unique as a species is that we are thinking, almost constantly, about what others think.

We think about what we ourselves think. And what others think we think. As the cognitive theorist Nicholas Humphrey has aptly put it, humans are the "natural psychologists" of the animal kingdom.

When someone else does something odd or surprising, our explanatory drive kicks into high gear. Having our social expectations violated (which is to say, when others act in ways that we're not anticipating or in a manner we failed to predict) unleashes a burning desire to uncover why they did that, and usually the only satisfying answer is one that involves getting into the actor's imagined headspace.* *Why did she slap me? Why is that man screaming about tacos? Why did the cat scale to the top of a giant conifer and meow its head off?* This sort of knee-jerk search for meaning doesn't always, or even usually, lead to correct explanations for why others do what they do. I might assume that she slapped me because I did something wrong—always my go-to assumption—whereby in reality she's mistaken me for the jerk who said something vulgar to her at her cousin's barbeque a few weekends ago. But it does always lead to some kind of mental state attribution.

In pondering the cause of an unusual death (or an unusual near-death, in the case of Tommy), this innate penchant for psychoanalyzing is on full display. "What were they thinking?" we ask. "What on earth was going through their head?"

In fact, I found myself asking this just the other day after reading a news report about two women who'd decided to take a late-night dip in the crocodile-infested waters of Daintree National Park in Queensland, Australia. One of these women, not surprisingly, was found a week later in the belly of a fourteen-foot estuarine croc. Not only were there signs warning of the presence of the world's largest carnivorous reptiles all over the beach where these two went frolicking, but they'd entered the water only a stone's throw away from the well-advertised base camp of a company offering daily crocodile-sighting boat tours. "You can't legislate against human stupidity,"

*So pervasive is our tendency to see deliberate minds in the world that we even find ourselves slapping and cursing at our misbehaving laptops or kicking the tires of our stupid broken-down cars.

said Warren Entsch, who represents the region as a member of the Australian Parliament. "If you go in swimming at 10 o'clock at night, you're going to get consumed." A bit harsh, perhaps. Really, though, what *were* they thinking? One would have to be foolish, suicidal, or, most likely in this case, very drunk to do such a thing. No matter your answer, however, and perhaps it's some combination of all three, you can see how our understanding of the situation rests entirely on our ability to deduce the dead person's intentions. A true solipsist will (rightly) point out that we can never be certain that another person has a mind at all, but let's assume that the victim did in fact have one and, given that it was fueled by a brain similar to the one presently idling in our own craniums, we're probably at least in the right ballpark with these mental state attributions.

How can we know that? Because if our own final words on this earth were "A croc's got me!" (as the victim's here reportedly were) while standing thigh-deep at night in a notorious crocodile habitat, these are the types of psychological factors we'd imagine would have led us to give this unfortunate parting line.

*

Anyone who has ever experienced the sudden, unexpected loss of a friend or loved one by suicide knows all too well the terrible need to make sense of the person's bewildering act. This is especially true for those who aren't left with a suicide note. To convince herself that her father really did want to die and didn't change his mind while it was happening, one young woman went so far as to place the noose around her neck and reenact the tragic event:

> All I could think about was what was he thinking at that very moment when he put the rope around his neck, when the last thing he saw was the family portrait in front of him. Did he think of us? Did he struggle? Did he do it, then try to pull out? So many questions. . . . [Personally] I know if I can't breathe I struggle and try anything to get air. . . . Some things I had to know. I put the chair in the exact place [where] my dad kicked it out from underneath. . . . I now know

> that he could have pulled himself back up by grabbing one of the steps. My dad did not want to stop or pull out. I get much more peace knowing he didn't want to get out and could, so this has helped me.

For ambiguous cases, coroners and forensic investigators must similarly determine whether the person died by suicide, homicide, or by accident, or perhaps even as the result of a *parasuicide,* which means that the individual didn't really mean to die by their self-harming behavior, but unfortunately they underestimated the lethality of their "cry for help."

To solve such mysteries, these experts attempt to channel the intent of the actor by, theoretically, entering the dead person's mind, much as the woman above did after her father's suicide. Strand by strand, investigators must peel away the murky layers of mental causation leading to the corpse they see before them. Did the person leave a note? How "serious" was the method used? (All else being equal, someone slitting their wrist or overdosing on drugs has a decent chance of living to see another day; for those who use a gun or hang themselves, by contrast, those odds drop sharply.) Did her death occur at a time or place she knew others were unlikely to find and rescue her? Was she on any medication? Was she under the influence of drugs or alcohol? Had she been depressed? Had she mentioned or threatened suicide in the past few days? Had she made previous suicide attempts? Did some major life event—a failed romance, problems in the workplace, financial troubles—have her unusually anxious or worried?

The technical term for this ability to get inside of someone else's head and (trying to) make sense of their behaviors is called a "theory of mind": because we can't literally see or feel other people's thoughts, emotions, or intentions, they are theoretical constructs to which we appeal in our attempts to understand, explain, and predict behaviors. Children develop a theory of mind around the age of three or four. And once we begin to process the social world this way, it's pretty much impossible to turn it off; hence, the human being as the animal kingdom's natural psychologist. This is also why the average preschooler is much more likely to be in the habit of asking her parents *why, why, why?* rather than *how, how, how?*

Having a theory of mind is a wonderful thing in that it allows us to empathize with our fellow earthlings. But it didn't come without some significant costs as well. One of these is the terrible burden of suicide bereavement. "Why, why, why?" we ask after such deaths. And all too often, there's just no satisfying answer. We're deprived of the explanatory catharsis we need so desperately. As Edwin Shneidman wrote, "The person who commits suicide puts his psychological skeleton in the survivor's emotional closet."

Sylvia Plath, the poster child of suicidal despair and creative genius, left a note of instruction with the words "call Dr. —" for whoever found her with her head stuck in the oven, a gesture that her friend the literary critic and poet Al Alvarez believes shows that she didn't really want to die.* Alas, it was too late.

But ironically, consider Plath's first suicide attempt ten years prior at the age of twenty, the incident that precipitated her lengthy stay in a mental institution and the period of her life that she chronicled in *The Bell Jar*. "[That earlier] suicide attempt . . . had been, in every sense, deadly serious," wrote Alvarez in his book *The Savage God*:

> She had carefully disguised the theft of the sleeping pills, left a misleading note to cover her tracks, and hidden herself in the darkest,

*Before sticking her head in the gas oven, Plath had sealed off the windows and doors of the kitchen with towels to prevent the toxic fumes from escaping into the rest of the house. It was early in the morning and her young children were still sleeping upstairs. If you believe Alvarez's somewhat circuitous line of reasoning, however, the poet timed her suicidal act so that it would coincide with the au pair's expected arrival that morning. Plath, he believes, anticipated that the girl would awaken the elderly landlord in the flat below with her pounding on the door. What Plath didn't factor in, however, was the fact that the old man slept without his hearing aid on, and also that he'd be knocked unconscious himself from the carbon monoxide seeping in from his own ceiling, since the famous poet's kitchen was directly above his bedroom. "The pity is not that there is a myth of Sylvia Plath," Alvarez laments, "but that the myth is not simply that of an enormously gifted poet whose death came carelessly, by mistake, and too soon."

most unused corner of a cellar, rearranging behind her the old fire logs she had disturbed, burying herself away like a skeleton in the nethermost family closet. Then she had swallowed a bottle of fifty sleeping pills. She was found late and by accident, and survived only by a miracle.

In clinical speak, the retrospective dissection of the victim's behaviors leading up to the death is called a *psychological autopsy*. The approach hasn't gone unchallenged, since no matter how hard we dig, whom we talk to, or what physical details of the scene we focus on, ultimately we're still seeing it through the lens of our own minds and can't help but make certain biased assumptions about someone else's stream of thought. But it does at least help to clear up that all-important question of whether the person intended to die.

Even this—deciphering intent—can be surprisingly tricky, though. When all you have in front of you is a naked dead body hanging from a rope in a closet, as the Bangkok police did in the case of seventy-two-year-old *Kung Fu* star David Carradine back in 2009, there can be, er, explanations other than suicide. A few murder conspiracy theories are still floating around, but they seem unlikely. Talks with the actor's former wives revealed his penchant for sadomasochistic sex, which, together with the cord allegedly wrapped around his genitals, led to the police ruling out suicide. It happens more often than you might think. The pathologists Riazul Imami and Miftah Kemal have estimated that based on records stretching back to 1791, each year over five hundred people lose their lives as the result of an episode of autoeroticism gone bad, the vast majority asphyxiations by ligature or hanging.

Other confounding cases involve those nebulous dream states of the parasomnias. What looks at first glance to be an "obvious" suicide can turn out, upon closer examination, to be a sad incident of "complex sleep-walking." In 1993 a twenty-one-year-old college student, clad only in his boxer shorts in the tundra-like cold of an early February morning in Iowa, quietly exited his apartment, jumped over some nearby overpass pillars, and started sprinting down the middle of a busy highway. A semitrailer struck and killed him. Initially classified as a suicide, the medical examiner changed the cause of the young man's death to "accidental death due to sleepwalking" once he learned of the victim's history of frequent sleepwalking and upon hearing that he'd been sleep-deprived (a known trigger for parasomnias) after several days of cramming for upcoming exams. His college roommate also shared the revealing fact that, just a few weeks earlier, the victim had spoken of having a recurring dream "in which he was running a foot race with someone from a nearby small town."

*

When the decedent in question isn't even human, understanding the motivating psychology of the actor poses special challenges. Still, that has never stopped people from finding suicidal intent in the deaths of animals. As far back as the second century, the Greek scholar Claudius Aelian wrote a whole book on the subject, filled with wild tales that today could be read as anthropomorphic fables.

Stories of animal suicide often sound like dramatic Orwellian fiction, particularly those from the early twentieth century. "The question as to whether animals commit suicide was settled to the satisfaction of a number of visitors at the Central Park Menagerie," reads a 1913 article from the *Washington Post*, "when a yellow-haired bulldog jumped into the open-air hippopotamus tank." The gobsmacked crowd could only watch in horror as one of the young hippos tossed the poor dog around like a rag doll. When the zookeeper finally fished the lifeless canine from the water, he concluded that "the motive for suicide was apparent . . . its emaciated condition showed that it was homeless and friendless."

In Geneva, New York, the same newspaper reported how an Alsatian (or German shepherd) that'd been left chained up alone on a porch hanged itself in despair after overhearing its beloved owner talking about going off to Germany to fight in the war. "The dog took particular interest in the [conversation], showing it by his movements and the manner in which he would cling close to his master. . . . [T]he dog thought it had lost its master and ended its life."

Meanwhile, over in New Jersey, a cat named Topsy leapt dramatically from the second-story window of a house and broke her neck after an agent from the Society for the Prevention of Cruelty to Animals took her kittens away.

In Chicago, a lovesick pet monkey named Hooligan hanged himself from a gas fixture after his nineteen-year-old owner left him behind when she moved out of her parents' home to get married. For weeks, Hooligan had wandered disconsolately through the house searching for his mistress—and he'd already tried to kill himself by downing half a bottle of chloroform.

Returning from a long voyage to the Orient, the British steamer M. S. *Dollar* pulled into port in San Francisco with the sad news that its own resident monkey, a wily capuchin named Bok, had committed suicide en route. Bok had just been punished for getting into some wet paint in the engine room. "The culprit was given a sharp reprimand in the shape of several slaps on the back. Suddenly he freed himself and, giving a last sorrowful look at his chastiser, he deliberately threw himself into space and fell to the stokehold below."

A wan cat "thin as a shadow" that had made its home at a Brooklyn police station jumped off a bridge over the East River, convinced that the officer to whom it was dearly attached would never return from a week-long vacation.

A belabored draft horse in the Yukon, weighted down by a backbreaking load, wheeled suddenly around and threw itself off a precipice after being whipped mercilessly by his cruel owner. "It was a clean and bold leap," said a man who saw the incident unfold, "made, I am sure, with a full realization of the results."

The list is long. Cornered stags bravely hurling themselves off cliffs rather than be torn apart by pursuing wolves. An old circus lion, crestfallen after losing a fight to a brash young rival in full embarrassing view of the female lions, strangling himself by jumping over a stage barrier while wearing his neck chain. A bereaved dog plunging headfirst into a well . . .

If nothing else, these types of stories reveal the breezy intuitiveness of Romanes's argument by analogy and, by contrast, the cognitive effort that's required to employ Lloyd Morgan's canon. In the pedestrian world where there's no particular reason *not* to anthropomorphize (unless we'd rather not know that the cheeseburger in front of us once had an emotional life), Romanes beats Morgan, every time. Why? Because again, Romanes's method simply articulates what already comes so easily to us natural psychologists. When you witness a lonely cat, one that you already know to be pining for its absent owner, standing on the Brooklyn Bridge and looking plaintively out into the distance, then see it jump, squalling, into the water below, the motive of suicide does spring to mind. More readily than, say, the notion that the cat got spooked out by an old radiator hose it mistook for a snake and instinctively jumped back, only to find itself in a real predicament.

These melodramatic turn-of-the-century accounts of animal suicide flourished in a particular historical context. They were appearing in the wake of Darwin's revolutionary ideas about man's true place in the natural world. *Surprise*, people were finally realizing, *we're animals too!* For the first time ever, there was an actual scientific reason to contemplate whether other organisms might suffer so greatly that, just like us, they can even be driven to suicide.

This reckoning of nonhuman consciousness was a long-overdue cultural achievement. Animal abuse and maltreatment had been rampant, and the prevailing view of other species simply as ownable objects without subjective experience did them no favors. And scientifically, too, our Victorian predecessors were on the right track. We now know that the neurobiological substrates of physical and emotional pain are ancient and contiguous across at least the vertebrate species. To deny the basic premise that other animals have complex inner lives is no more reasonable than concluding that every human being but you is an unthinking automaton. When your dog sulks in the corner as he's watching you pack your suitcase for a two-day conference—I'm saying this from personal guilt-ridden experience—you're not being "fooled" into thinking that he's genuinely upset. He's seen this sequence of events before, he knows what's to come, and his emotions are real.

Likewise, when Tommy curls up with me each night, purring loudly and still kneading his claws into my flesh (our nightly ritual for the past fourteen years), there's obviously a complex brain in that head of his producing a warm glow of contentedness, one accompanied, of course, by an entire feline cosmos of who-knows-what. Since unfortunately I don't come equipped with night vision, heat-seeking whiskers, or ultrasonic ear canals, I'll never know what it's like to be a cat. I do have an idea of what raw happiness feels like, though, and whatever scientific or philosophical pathway I use to get there, I'm standing on pretty solid ground in concluding that Tommy's in the throes of it. After all, for both cats and humans, the act of cuddling releases exactly the same calming hormone—oxytocin—known to play a central facilitating role in the bonding behaviors of most social mammals.

Still, that's an altogether different thing than concluding that as a confused young cat with serious early abandonment issues, Tommy once climbed to the top of a six-story pine tree in an Arkansas forest intent on ending his angst-ridden life.

✷

To some, of course, the skepticism I've so far expressed about animal suicide simply betrays my stubbornly arrogant endorsement of what Ramsden and Wilson call "human exceptionalism": "When people reject the possibility of an animal committing suicide," those authors argue, "they reserve not only the act itself for humans, but many traits that enable it—emotion, intelligence, mind and consciousness. As a distinctively 'human privilege,' suicide becomes constitutive of the human mind." In other words, if you fail to accept that other animals—feeling hopeless, friendless, mistreated by humans, or depressed—can deliberately choose to terminate their existence, then you're just another one of those cold-hearted science types who, despite their apparent disavowal of any religious ideology, secretly believe that humans are at the top of the *scala naturae,* right beneath angels and rungs and rungs above all those grunting, soulless subhuman beasts. You're basically painted into the same unflattering, animal-hating portrait as C. Lloyd Morgan, our cynical old scorpion-torturer.

I really do understand the underlying sentiment. If it's not obvious already, I'm sympathetic to animal rights causes. But, to begin with, the logic is simply wrong. To question the existence of suicide in nonhuman animals does not equate to denying them "emotion, intelligence, mind and consciousness." Again, knowing what we do about brain evolution, and with recent gene-mapping technologies illuminating just how similar we really are to other organisms, it would be ridiculous to argue that these capacities are found in humans alone. But these aren't all-or-nothing traits, either.

As an undergraduate student, I spent several years volunteering at a great ape sanctuary in south Florida, which essentially meant babysitting a motley crew of toddler-aged chimpanzees and orangutans. Without fail, some visitor to the sanctuary would ask me, "So which are smarter, chimps or orangs?" What the person was really asking, of course, was which of these two species has an intelligence most like human intelligence. In fact, not only are we genetically more similar to chimps than chimps are to orangutans, but we're closer to chimps than rats are to mice. In the field of comparative psychology, there are really only different *kinds* of intelligence, not degrees. Every modern animal mind is adapted to the ancestral conditions under which it evolved. As a species, the way our brains operate today—the nature of our emotions, our sensory abilities, our critical-thinking skills, even our very experience of consciousness—was originally forged to solve problems encountered by our ancestors long ago. We'll explore some of those "problems" in the next chapter, but what's important to take away now is that something as simple as planning to buy avocados at the grocery store reflects a mind designed to plan for future needs. Maybe we're not always good at *remembering* to pick up avocados after work, but we easily project the self into a future where one's state of hunger will be different than it is now (in this case, a future that includes a hankering for guacamole).

When it comes to suicide, our evolved mental abilities are relevant when we consider the specific kinds of emotions usually involved. For instance, unless you're a psychopath, nobody ever had to sit you down as a child and carefully instruct you on how to go about experiencing the crippling emotion of shame, which is a well-known factor in many suicides. Like pride, shame is a self-referential emotion, a special subgroup of affect that hinges on our understanding—or at least our belief—that we're the focus of someone else's morally evaluative attention. (Compare this to more primal emotions such as happiness, anger, and sadness, all of which can occur without taking the perspective of another person.) Remember that we're natural psychologists. More often than not, we're thinking about what others are thinking. But here's the rub: this also means, for better or worse, that we're thinking about what others are thinking *about us*. And that's, I believe, a distinctively human emotional experience.

I wasn't joking about that psychopath thing, by the way. The psychiatrist Hervey Cleckley, the "father of psychopaths"—research on, that is—believed true psychopaths rarely, if ever, die by suicide because they lack the interpersonal emotions (and the conscience) necessary to drive them to it.*

✷

The hypothesis that suicide is a consequence of our species' emotional Achilles' heel finds support in a recent study by the neuropsychiatrist Martin Brüne and his colleagues. In comparing the brains of suicide victims to those who died of natural causes, Brüne zeroed in on an enigmatic spindle-shaped cell type called von Economo neurons, or VENs (named after their Austrian discoverer, the late and

*Subsequent research suggests a slightly more complicated picture of the relationship between psychopathy and suicidality than Cleckley depicted, however. Some investigators, such as the clinical psychologist Edelyn Verona, have found a positive correlation between suicidality (ideation and previous attempts) and the psychopathic traits of impulsiveness and aggressiveness. However, because these subsequent studies have relied mostly on self-reports by incarcerated psychopaths, and have failed to ascertain the seriousness of the methods used, it's impossible to gauge the actual suicidal intent of the inmates. Findings from a 1980 study by Michael Garvey and Frank Spoden, by contrast, seem to offer support for Cleckley's initial argument. In a sample of psychiatric inpatients, these authors found that although "sociopaths" did sometimes have a history of suicide attempts rarely were they serious. Of sixty-three suicide attempts by these individuals, only three were serious enough to require at least a two-day medical hospitalization. "The almost total lack of seriousness of the 63 attempts," write Garvey and Spoden, ". . . suggest that sociopaths have no real intention of killing themselves. The findings in our study would support the notion that sociopaths use suicide attempts to manipulate others or to act out their frustrations."

imposingly named Constantin von Economo). The brain scientists had a hunch that these distinctive neurons might somehow play a role in suicidal behavior. Although not reserved entirely for humans (they're also found in great apes, cetaceans, and elephants, notably all complex social species), VENs are significantly larger and more plentiful in human brains. In the fronto-insular cortex alone—which is basically the center of operations for self-awareness, empathy, and any other complex social cognitive function—a human adult has about 82,855 of these spindle cells, whereas the average chimp has a paltry 1,808.

The exact role of VENs is still shrouded in mystery, but these numbers alone betray the critical importance of perspective-taking to our species' adaptive success, given their remarkably recent evolutionary explosion on a scale of perhaps only the last 100,000 years. In Brune's comparison of suicidal and non-suicidal brains, the scientists homed in on the anterior cingulate cortex, a part of the brain devoted to processing complex negative emotions such as shame, guilt, hopelessness, and self-criticism, and where VENs are also found in abundance in human beings. Here, they discovered significantly greater densities of VENs in the brains of the suicide victims compared to those in the control group. "VENs, it seems," the authors of the study conclude, "are part of a neural circuitry subserving the highest cognitive functions that emerged during human evolution, which entails, perhaps inevitably, the potential of suicidal behavior."

Brüne and his team were aware of the potential confounding variable that the brains of those who die by suicide might show physical signs of their disorder. So to rule out the possibility that any differences found in VEN densities in the suicide brains were the result of a comorbid psychiatric condition, not suicidality, specifically, *all* of the brains had come from people diagnosed with either schizophrenia or bipolar disorder. That is, the entire collection of brains sampled for this study were from people suffering from mental illness; the decedents had just died of different causes (suicide or not). Still, that key VEN effect panned out: it was the brains of those who killed themselves that had the densely packed spindle cells.

"The ability to reflect upon oneself in ways that lead to negative self-appraisal, self-derogation, and feelings of shame, guilt and hopelessness" surmise the authors, "may put patients with psychotic disorders at risk of committing suicide." In other words, perhaps we've been looking at things the wrong way all along. Mental illness alone doesn't cause suicide. Some deeply troubled individuals are perfectly safe in their delusions. It's being aware that you're mentally ill, and believing judgmental others know this about you too, that makes mentally ill people so vulnerable to taking their own lives. This helps us to make sense of that otherwise perplexing clinical finding, known for some time, that patients who are aware that their delusions *are* delusions are ironically at increased risk of killing themselves.

✷

Even if other species lack the social-cognitive hardware needed for self-referential emotions such as shame, that doesn't mean they aren't conscious or don't have emotions or don't suffer as deeply, or even more deeply, or in different ways, than we do. It just makes them lucky not to have to undergo this torment of others' eyes on them, judging them as socially tainted individuals. As anyone who has felt even a taste of this emotion knows, shame can be unbearably painful.

What leads people in one culture to feel ashamed can be very different from what has folks wanting to crawl under a rock and die in another. That's where social learning comes in. If you grew up immersed in, say, the Bushido tradition of a fourteenth-century Japanese samurai, there were many reasons why you might have sliced open your own belly and eviscerated yourself in the symbolic (and gory) public act of *seppuku*. But just as it was for an eighteenth-century Afghan princess poisoning herself, a nineteenth-century British fishmonger hanging himself, or a twenty-first-century American tech executive jumping off the Golden Gate Bridge, chances are shame would have had something to do with it.

"Shame was the most terrible injury a samurai could contemplate," wrote the Confucian scholar Saitō Setsudō, "and he would rather die than suffer it. If charged with a crime and sentenced to death, he considered it an honor to be permitted to cut his stomach. He could not countenance the thought of being bound with rope—he was better off being a corpse." Compare that with the self-loathing lines from what some scholars believe to be the oldest suicide note in existence, a poem found scrawled by some unidentified man on a 4,000-year-old papyrus roll dating back to ancient Egypt's Middle Kingdom. "Lo, my name is abhorred . . ." begins each of the first three verses:

Lo, my name is abhorred,
Lo, more than the odor of carrion
On summer days when the sun is hot.

Lo, my name is abhorred,
Lo, more than the odor of crocodiles,
More than sitting under the bank of crocodiles.

Or consider Jocasta's suicide in Sophocles' *Oedipus Rex*. She's prepared to live with the unsettling knowledge that her husband—and the father of four of her children—is in fact her adult son, so long as no one else knows the truth. It's only when Oedipus starts digging around and threatens to get to the incestuous bottom of things that Jocasta hangs herself. She kills herself not out of guilt for her actions, but from the shame of others finding out.

Jean-Paul Sartre knew the searing effects of shame on the human mind too. In his play *No Exit* (where that oft-quoted line "Hell is other people" comes from), Sartre tells the story of three recently dead strangers who find themselves locked up together for all eternity in a single windowless room, a place where their shameful earthly sins are slowly, scandalously exposed and forever judged by one another. The condemned can never fall asleep; the light is always on—even their eyelids are paralyzed so that they can't blink to avoid the others' glares. The play is an allegory that forces us to examine the subtle ways by which other people, through their sheer presence, their knowing gaze, can cause us psychological distress. And suicide? That's off the table too, given that they're already dead.

"Open the door!" one of the characters pleads to the menacing demons just outside.

> Open, blast you! I'll endure anything, your red-hot tongs and molten lead, your racks and prongs and garrotes—all your fiendish gadgets, everything that burns and tears and flays—I'll put up with any torture you impose. Anything, anything would be better than this agony of the mind, this creeping pain that gnaws and fumbles and caresses one and never hurts quite enough.

I'm not sure about you, but that kind of suffering doesn't sound like a "privilege" or something I'd wish on any animal I know.

It's not always simply feeling others' fiery gaze on us, of course. Sometimes we suffer from the opposite problem. The absence of being gazed on, of being recognized as a valuable member of society or deserving of protection and happiness, can be just as devastating for a member of a species whose deepest emotions are so tightly bound to others' thoughts. In short, we need others to think that we matter. The developmental psychologist Philippe Rochat calls this "the basic affiliative need":

> We essentially live through the eyes of others. To be human . . . is primarily to care about how much empathy, hence acknowledgement and recognition of our own person, we generate in others—we care about our reputation as no other animal species does.

Here's how one middle-aged woman, who'd endured years of sexual abuse as a child without her mother ever stepping in to intervene, explained her ongoing suicidal feelings to my PhD student Bonnie Scarth, who was conducting interviews for a thesis on the subject:

> It is just so hard. I want to forgive my mum and I want to be able to move on, and those are things I can't resolve. She is dead. How can I make her see what she did? You know? How she left me feeling? . . . I can feel like I matter or like I don't. I think a lot of people who

are feeling suicidal often feel that way, like you don't matter and you are not important. That is what puts you in the mood.

*

On the matter of animal suicide, where does all of this leave us then?*

There's no question that chimps, our closest relatives—and therefore the species we'd expect to find the most behavioral similarities with our own species—lead rich social lives brimming over with emotional turmoil. Yet over the course of what amounts to centuries of meticulous observation by primatologists at hundreds of different sites, on no occasion has a distraught or ostracized ape ever been seen, for example, to climb to the highest branch it could find and jump.

That's us. *We're* the ape that jumps.

There will always be intelligent people who swear they've seen an animal intentionally end its own life and can think of no other explanation for the behavior. To say "prove it" is simply to invite the perfectly reasonable response "prove that it's not"; so given those interspecies boundaries between minds discussed earlier, there's not much of a conversation to be had if that's a person's firm conclusion.

Yes, there's that old mantra in science: absence of evidence is not evidence of absence. I've always hated that saying. Not because it's wrong (it's absolutely not), but because it makes a plausible, data-derived argument (for instance, the argument that suicide doesn't occur in other species, or that there is in fact *not* a tiny invisible Ferris wheel filled with little pink poodles spinning atop your head right now) seem like it's on equal footing with the highly improbable.

*The animal behaviorist Marc Bekoff concludes, cautiously, that "it's too early to make any definite statements about whether other animals commit suicide," after relaying a story told to him by an audience member following one of his talks—a tale about a donkey that purportedly walked into a lake and drowned after giving birth to a stillborn deformed foal. Antoni Preti, a psychiatrist who reviewed forty years of peer-reviewed animal ethologies in search of credible accounts of suicide in the wild, disagrees. "Naturalists have not identified suicide in nonhuman species in field situations despite intensive study of thousands of animal species," Preti argues.

So let's analyze the empirical evidence. That, after all, is the bedrock of good science. If there may be other explanations for an animal's self-destructive act than suicide, we shouldn't let our tendency to anthropomorphize prevent us from finding them. We can't go back in time to the wistful days of the Victorian era, when suicide among pets seemed to be in vogue, but sensationalized reports still occasionally make the rounds. Those we can, and should, scrutinize.

Consider the myth of the leaping lemmings. Throughout much of the last century, the public was led to believe that lemmings routinely jump off cliffs in a zombie-like ritual of mass suicide, one after the other dropping into the Arctic Ocean, as shown clearly in the 1958 Disney documentary *White Wilderness*. Later it was revealed, however, that the story was concocted by the producers of the film, who'd staged the famous death-plunge scene by herding the animals to the cliff's edge and then launching them, perversely, into a river near downtown Calgary using a makeshift turntable.

A more recent example is the strange case of Overtoun Bridge near the banks of the river Clyde in Scotland. Since at least the 1950s, a curious number of unleashed dogs out on leisurely strolls with their owners have made a beeline to a specific area of the bridge's ashlar parapet, only to jump over the wall and come crashing down into the rocky stream fifty feet below. Once the press picked up the story of the "Dog Suicide Bridge" of Dumbarton and it began to circulate online, a canine psychologist and animal habitat expert teamed up to investigate using controlled experiments. The researchers concluded that the "suicidal" dogs—nearly all of which were long-snouted scent hounds—had most likely been drawn to the strong odor of male mink urine wafting up from a nesting site in some underbrush beneath that part of the bridge. Mice and squirrels were also discovered in the area. According to David Sands, the canine psychologist of the duo, the features of the rampart, combined with the enticing scent of ambient prey, drove these dogs, literally, over the edge. "When you get down to a dog's level, the solid granite of the bridge's 18-inch thick walls obscures their vision and blocks out all sound," Sands told the *Daily Mail* back in 2006. "As a result, the one sense not obscured, that of smell, goes into overdrive."

Sheep have made an appearance in recent media claims of animal suicide, too. In 2005 a mild-mannered shepherd having breakfast in the small village of Gevaş in eastern Turkey, was horrified to watch as his large flock of sheep, which moments earlier had been grazing quietly in the distance, suddenly darted off to a nearby cliff and flung themselves, one by one, into the abyss. Upon arriving at the gruesome scene, first responders discovered that, well, there was good news and bad news. The bad news was staggering: nearly 1,500 sheep had hurled themselves off the precipice. The good news? Only 450 of the animals actually perished that day. "Those who jumped later were saved as the billowy white pile got higher and the fall more cushioned," a Turkish newspaper reported.

It's unclear what set this bizarre chain of events into action (in addition to the tragic deaths of the sheep, the economic livelihood of the already poor village was affected), but a similar case of 400 "suicidal sheep" who'd thrown themselves into a steep ravine in the French Alps a few years before the Turkish affair might shed some light. Investigators there determined that the mostly young sheep had been startled by a pack of wolves (which had controversially just been reintroduced into the region) and the entire flock thrown into a sort of contagious panic. When triggered by predators, a violent storm, or some other imminent threat, it's easy to see how herd behavior can be deadly. All it takes is a few terrified individuals running off blindly to escape and in the wrong direction.

Whale beachings have also featured in the ongoing debate over nonhuman suicide. Accounting for about 2,000 cetacean deaths per year, these animals seem to come inexplicably ashore, typically many of them together and from the same tightly knit social pod, only to get stuck on their sides and die slowly in the sun from stress, dehydration, and the incoming tide filling up their blowholes. Although this baffling behavior used to be referred to as "intentional stranding" by experts, most marine biologists today tend to favor hypotheses that treat beaching as an unfortunate, but unintentional, by-product of other causes. Contenders include freak weather occurrences, human disturbance, or the intense social bonds in some cetacean species leading to mass strandings. Studies have confirmed that the use of military sonar, for example, can cause beachings by interfering with the animals' bioacoustics, leading them to make fatal navigation errors. Other research findings suggest that the herdlike social behaviors in some whale species factor into the phenomenon. Strandings might occur in situations where, for instance, a sick or injured whale goes to coastal shallows and the others follow suit, or even when large cetaceans unwittingly trail smaller, speedier dolphins into hazardous conditions after they proved to be reliable guides to feeding pools in deeper waters.

Of course, any given whale stranding could be a combination of these or other variables. Certain coastal regions, such as Farewell Spit on the northern coast of New Zealand's South Island and Geographe Bay in southwestern Australia, have developed reputations for their frequent mass strandings. What these hot zones share in common is a distinctive coastal topography, specifically, a protruding section of coastline adjacent to gently sloping sandy beaches . . . so gently sloping that echoes are perilously muted for an animal dependent on its echolocation to navigate its surrounds and avoid the reefs.

In recent years, the oddly self-destructive behaviors of animals infected by parasites have also received considerable attention under the inaccurate heading of suicide. These cases should be filed under "invasions of the body snatchers." The most commonly cited example is that of rodents that have picked up the *Toxoplasma gondii* parasite, a protozoan causing the animal to lose its fear of the scent of cats, which you'll probably agree is a pretty useful instinct to have for a rodent.*

My own favorite is the tale of the wood cricket and the horsehair worm, the latter of which looks just like it sounds it should. As soon

*Biologists Midori Tanaka and Dennis Kinney put forward an intriguing hypothesis in 2011 suggesting that human beings infected with the *T. gondii* parasite are at increased risk of suicide. The authors cite the comparatively high suicide rates among individuals such as veterinarians, waitresses, nurses, and farmers, whose jobs place them at greater risk of exposure to *T. gondii* and other infectious diseases. After controlling for other possible interpretations for the correlation, Tanaka and Kinney suggest that it probably has something to do with the parasite kick-starting mental disorders, such as depression and schizophrenia, among those already predisposed.

as these worms wend their wormy way into the body of a cricket, they set about hijacking the latter's central nervous system, then proceed to steer their mindless host to the nearest body of water. Once there, the cricket—which normally avoids water, since it can't swim—hops right in and drowns. The worms slither out of its dead head, then cavort and reproduce in the water; their larvae are devoured by mosquitoes or mayflies scavenging the water's edge. Those flying insects eventually leave shore and in turn get snatched up by a hapless wood cricket and, well, it's the circle of life.

The defining element of suicide is, again, the intention to die by a self-directed fatal act. When a suspected suicide occurs, the incident should somehow reveal this unambiguous intention. By this standard, evidence for suicide in other species is indeed strikingly absent. As it stands, the only clear takeaway message from most stories of animal suicide is that too many people are incompetent caregivers, such as the surprising number of pet owners who apparently think it a good idea to leave their charges on raised platforms, tied up by long neck chains.

*

For scholars who see suicide more as the end point on a continuum that includes sub-lethal acts of self-harming, dangerous risk-taking behaviors, catatonic depression, and anorexia, the question becomes more complicated. Nonhuman animals kept in abysmal captive conditions of isolation or overcrowding, for instance, often display disturbing patterns of self-harm, severe stress, and symptoms of mental illness. Parrots may pluck themselves bald, tigers pace restlessly in their cages, lab monkeys rock back and forth in a trancelike state, morose whales float listlessly in their theme-park tanks, and so on. In the wild, some juvenile chimps, such as eight-year-old "Flint" whose sad case was so famously documented by Jane Goodall back in the 1970s, can apparently become so depressed and lethargic at the deaths of their mothers that they stop eating and eventually starve to death themselves. There are similar cases of anorexia-related fatalities in depressed dogs mourning the deaths of their owners and refusing to take food from any other human.

A more liberal conception of suicide makes science writer Laurel Braitman hesitant to say that the behavior is uniquely human. "Nonhuman animals also have their own continuum of self-destruction," she writes in her book *Animal Madness*. "They may have fewer tools available to them to inflict mortal wounds and also lack humanity's sophisticated cognitive abilities with which to plan their own ends, but they can and do harm themselves. Sometimes they die." There's not much to disagree with in that assessment. But it also doesn't tell us anything especially helpful when it comes to addressing whether other species ever act with intent to end their own lives, the sine qua non of suicide.

In 1897 Émile Durkheim, the renowned French sociologist, published his groundbreaking book *Suicide*, a work showing how certain telltale social demographic data such as religiosity, unemployment rates, and marital status can be used by researchers and clinicians to get a better handle on a nation's suicide trends, helping to predict ebbs and flows in this alarming behavior at the population level. Although Durkheim's model didn't focus on the nuts and bolts of psychological causation, he nevertheless rejected the many anecdotal cases of nonhuman suicide that he'd come across. Dogs perishing from starvation after the death of their masters? That's only because, Durkheim reasons in his introduction to *Suicide*, "the sadness into which they are thrown has automatically caused lack of hunger; death has resulted, but without having been foreseen . . . the special characteristics of suicide by us are defined as lacking."

I'm with Durkheim. In my opinion, suicide is like pregnancy. Just as it's illogical to speak of being "a little bit pregnant," viewing suicide in terms of varying degrees makes little sense. With intention at the core, either a death is or is not suicide.

✷

At first blush, drilling into the seemingly implacable question of whether suicide is a uniquely human behavior may appear irrelevant—or at least unnecessary—when it comes to reducing the number of human suicides. It would be if we stopped there, merely stating that an evenhanded review of the literature strongly suggests that, yes, it is uniquely human. But upon arriving at this conclusion, an even more important set of questions comes into focus: If suicide is uniquely human, motivated by uniquely human emotions, is it simply an unfortunate by-product of our mental evolution, or is it possible, strange as it may sound, that suicide actually evolved as an adaptive program of self-destruction? Without waking the ghost of Freud, could it be that there really is such a thing as a latent "death drive" in all of us, one that's set into motion under emergency social conditions in which killing yourself has (or once had) the counter-intuitive effect of improving the odds of your genetic survival? Some experts have made that surprising claim. Other researchers disagree adamantly, arguing instead that suicidal thinking is and always has been pathological, a clear sign of a defective and diseased mind.

It's to this fierce debate we turn our attention next. And the stakes are high. Because if the former are correct, and suicidality really was designed by nature to perform a specific job in our species, reverse-engineering this elusive design may help enable us to defuse the suicidal brain's default wiring. As any good saboteur will tell you, you can't disrupt the functioning of a device until you know what it's meant to do.

3

betting odds

Sometimes it's better to light a flamethrower than curse the darkness.

Terry Pratchett, *Men at Arms* (1993)

On the bright early morning of April 20, 1965, fourteen-year-old Denys deCatanzaro opened his eyes to find that his father, whom he'd been expecting to wake him up early for church, their normal Saturday morning routine, had already left without him. Looking out the window, he caught a glimpse of the family vehicle driving mysteriously away.

"Saturday mornings were special for us," Denys explains. "We'd drive together into downtown Chicago to a little Episcopal convent where I'd serve as an altar boy while he'd say mass to a small group of nuns. I awoke to find that he'd apparently gone on his own."

There was an eerie silence in the house, broken only by what sounded like the muted sobs of his mother coming from down the hall. Something was wrong.

Denys walked with trepidation into the bedroom of his eleven-year-old twin brothers. "What's going on?" he asked his mom. "What is it?"

"Gregory." That was all she could say. Gregory.

His older brother was dead.

"What a shock," Denys recalls more than a half-century later.

"That really threw me for a loop. I mean, that sent me into depression, the only time in my life, I think."

Prior to this tidal moment, Denys, like most young teenagers, hadn't given much thought to his sibling's state of mind. All he knew, really, was that nineteen-year-old Gregory, the eldest of six from an upper-middle-class home, had been studying at the University of King's College in Halifax, Nova Scotia, some 1,600 miles away from the family's residence in Evanston, Illinois.

"He was expected to be home within days for the summer," said deCatanzaro, now a biology professor at McMaster University. "From what I understand, he went onto a bridge over some railway tracks and jumped."

I cringed, noticeably, on hearing this. "Yeah, grisly," said Denys.

The boys' father, Carmino Joseph deCatanzaro, who went by "Bruno," was a tall, composed man more Scandinavian in appearance than Italian. A serious scholar with sincere religious convictions, Bruno had a PhD in theology and worked as a cleric while teaching part-time at the University of Toronto. Denys's mom, Joan, raised the children.

Six years before Gregory's suicide, however, the deCatanzaros had moved to Illinois, where Bruno had accepted a job as an associate professor at an Episcopal seminary. At that time, firstborn Gregory, a precociously bright thirteen-year-old, much like his studious father had been, was still enrolled at Upper Canada College, an elite Toronto boarding school just up the road from the house. It was a difficult decision, but together the family decided it was best to leave Gregory there, ensconced in his studies.

"He was largely on his own since his early teens, which in hindsight was probably a mistake," Denys said.* "He was emotional and sometimes impulsive as a child."

*Gregory's absence was almost certainly on his brother's mind when Denys noted in an article that "suicide often occurs when an individual has little contact with his family; in fact, protracted isolation from family and other members of society is among the best predictors."

Soon enough, though, Gregory would continue on to pursue his bachelor's degree at Halifax. And as far as anyone knew, all seemed to be going okay.

*

Navigating the suicide of a loved one can drive some people further into the comforting embrace of religion. The calm whisper of an ultimate reason or purpose to such a painfully inscrutable experience, even if we mere mortals can never know, offers a reprieve from the constant questioning and the waves of self-recrimination and regret. For Bruno, a gentle, private man who'd always been more comfortable having spirited conversations about history and linguistics than he was speaking about difficult personal matters, Gregory's suicide only cemented his faith. "My father was deeply confused by my brother's death, and it seemed to make him even more religious," said Denys.

Shortly after the tragic loss, the wounded family returned to Canada. Bruno had decided, "perhaps impulsively," Denys surmised, to take a job back in Ontario as a parish priest. In 1980 Bruno would be named bishop of the Anglican Catholic Church of Canada. He died of a massive stroke three years after that.

For Denys, Gregory's death had the opposite effect. "What did it do to my religion? It knocked it off its foundations," he said. Instead, Denys found himself gravitating to the natural sciences in search of answers. As a junior in high school, while his father was just beginning to translate the first English version of the Gnostic Gospel According to Philip from the Coptic, Denys read the zoologist Desmond Morris's controversial book, *The Naked Ape*. "It shook things up in that era and got me interested in evolutionary issues."

It's easy to see why Morris's book would have appealed to any healthy sixteen-year-old boy. Among its more scandalous content, it included rogue theories about the evolution of man's comparatively large penis in the animal kingdom, women's perennially swollen breasts, and the earlobe as a special catalyst for orgasm. It was in fact so salacious for its time that the book was removed from the shelves of several U.S. libraries.

But look a bit deeper and you'll also see how such a book would have offered something more than titillation to someone like Denys, a shy, intelligent teenager still dealing with the emotional aftermath of his brother's suicide and receiving only well-meaning but unsatisfying religious answers at home. Whether Morris's often fanciful theories of human nature were accurate or not, his mere framing of our species, in all its urbanized, perfumed glory, as a type of peculiar hairless primate would have been nothing short of an epiphany to an adolescent in the midst of an existential crisis. Like any animal, our long-ago ancestors were shaped by the shifting sands of natural forces, and if only we trained our eyes just right, Morris told his readers in *The Naked Ape,* invisible lines of order pointing to our place in the universe could be made out in the clarifying light of science.

At a time when biologists subscribed whole cloth to Darwin's basic survival-of-the-fittest model, however, the question of how suicide could possibly be explained by evolution was, well, something of a problem. Nature "will never produce in a being anything injurious to itself, for natural selection acts solely by and for the good of each." Those were Darwin's exact words in *On the Origin of Species.* How could it possibly work any other way? Even Morris, who had a suspiciously clever evolutionary theory for just about everything, was just like every other adaptationist of his day when it came to suicide: he had nothing at all to say.

Still now it appears, at least at first blush, rather absurd to ponder if such an obviously self-defeating behavior as suicide might have evolved for an adaptive purpose. By taking your own life, you're removing yourself from the evolutionary game and taking your reproductive potential, or whatever is left of it, with you. Isn't that the antithesis of adaptive? But maybe we just need to adjust the light a little. That's what Denys found himself slowly beginning to do about a decade after his brother's suicide.

*

To speak of the timing of such a paradigm shift as serendipitous feels somewhat impolite, but nonetheless, Denys's attempts to make sense of Gregory's death from an evolutionary perspective were aided by a convenient revolution unfolding around this time.

It began in 1964, when a young British polymath named William D. Hamilton published a landmark theory on what came to be referred to as Hamilton's rule. What his insight did, in essence, was work out a long-standing kink in Darwin's almost biblical treatise that had been pestering biologists for nearly a century: if nature red in tooth and claw is simply about survival of the fittest, then why do many social animals sacrifice their own interests to ensure the survival not of themselves, but of others?

Examples abound. Rather than go off to breed on their own, for instance, juvenile male woodpeckers sometimes stay in the nest to help their parents raise younger hatchlings, and asexual worker ants labor tirelessly as drones and foragers to ensure their queen is well nourished and her offspring survive. Hamilton managed to puzzle together a mathematical formula that successfully predicted the likelihood of such cases of biological altruism (helping another despite one's own reproductive interests being compromised).

Hamilton developed the formal algebraic model of kin selection: $r \times B > C$. First, there's the *r* component. This stands for *relatedness*—specifically, the genetic relatedness between two organisms. When comparing biological kin—say, your sister versus your first cousin—you share more identical genes with the former (~50 percent) than the latter (~13 percent), and so the value for *r* is greater for your sister. Second, there's the *B* in Hamilton's equation, which stands for the reproductive *benefit* to this other individual. Finally, the C in this equation is the reproductive *cost* to you, our prospective altruist. The plug-in possibilities for the equation, at least when it comes to human social behavior, are nearly limitless, but presumably, theoretically, they're still calculable. When benefits are measured this way, in the currency of genes, altruism is most likely to occur whenever the benefits of helping one's kin exceed the costs to the self.

The stochastic gains of kin selection apply to any social species. Staying in the nest to help its parents raise eight additional hatchlings is a smart tactic for a young male woodpecker who'd probably only have two or three offspring of his own; and since the queen is the mother of all the rest, what's good for the brood is good for the lowly worker ant, too. This new gene-centric view of evolution, with its emphasis on kin selection, being ushered in by Hamilton and a few other leading thinkers of the mid-1960s was a critical addendum to Darwin's original model of natural selection, and it was one that held clear implications for understanding our own species.

In 1975 such a perspective materialized into E. O. Wilson's classic book *Sociobiology: The New Synthesis,* and a year after that, it was further popularized by Richard Dawkins in *The Selfish Gene,* which showed how so much of human behavior, just like the behavior of every other animal, is in fact a sort of illusory puppetry. Dawkins argued that, in spite of our conscious interpretation of our own motives, we act the way we do mostly to maximize the chances of transferring as many duplicates of our genes as possible to the next generation. Most of these genes tend to get out there the old-fashioned way, through direct sexual reproduction. But as Hamilton's rule dictates, genetic immortality can also be attained indirectly by mechanisms such as kin selection, in which we aid our relatives. As the population geneticist J. B. S. Haldane cheekily responded on being asked if he would give his life to save his own drowning brother: "No, but I would to save two brothers or eight cousins."

There are other forms of biological altruism as well, including tit-for-tat strategies (or reciprocal altruism) by unrelated individuals, such as grooming in baboons who take turns picking beasties from the other's back; and potluck meals in vampire bats, who graciously regurgitate food into the mouths of hungry friends after a successful hunt. But in the end, it all boils down to cumulative genetic success.

This tidy articulation of love and hate and other affairs of the human heart using the mathematical principles of inclusive fitness—an ever-changing algorithm reflecting the proportion of your DNA that will outlast your ephemeral presence as a gene-shuttling transporter here on earth—soon solidified into the bedrock theories of sociobiology and evolutionary psychology, disciplines emerging at the height of the disco years.

✵

For Denys, who was by this time a bell-bottom-wearing master's student at Carleton University, these academic developments were fascinating. But it wasn't until he found himself face-to-face with some patently *maladaptive* examples that it all began to click.

One summer Denys was an intern at a hospital in Kingston, Ontario, "one of those old institutions," he told me. "Now all of them are shut down, pretty much—but it had well over a thousand residents. I was assigned to work with very disturbed kids who were head-hitters, self-scratchers."

He started reading as much as he could about what looked to him like extreme emotionally driven self-harming.

"It had been nearly ten years," he recalled, "but the experience put Gregory's suicide back at the front of my mind. There are so many cases where it's emotion that's driving suicide. Intense emotion. Well, I guess the idea for my 'mathematical model of self-destruction and preservation' grew from there."

The title may be a mouthful, but Denys's theory of suicide, the first of its kind to apply evolutionary reasoning, is actually fairly intuitive. In a nutshell, it suggests that we're most likely to die by suicide when our direct reproductive prospects—having (more) children—are discouraging and, simultaneously, when remaining alive threatens our genetic success by impeding or stopping altogether that of our close biological kin. There are many variations of the theme, but in my mind's eye, this conjures up the archetypal image of the stubble-faced loser who just can't seem to get his act together and owes a small fortune in gambling debts, who survives day to day by leeching off his more successful brother and draining resources that'd otherwise go to the latter's adorable kids.

In any event, Denys first laid out his ideas in 1980 in a lengthy theoretical article published in *Behavioral and Brain Sciences*, a piece titled "Human Suicide: A Biological Perspective." (That was the same year, incidentally, that his father, Bruno, became bishop.) Here, in Denys's own words from that seminal article, is how he saw human suicide fitting squarely into evolutionary theory:

> If an individual's present and future behavior is unlikely to change the status of his genes, there may be no ecological pressures preventing his suicide. If an individual is unlikely to reproduce, unable to support himself and his family adequately, and unable to contribute to the welfare of other reproducing individuals sharing his genes, his death may not affect the frequency of genes he carries. Suicide would consequently not eliminate any genes from the gene pool that were not already eliminated. Thus, under the limited ecological conditions in which suicide appears to occur, there may be no selective pressures to prevent it. . . . [This could] operate against one's survival when one consumed resources without being productive. Where there is "no reason to live," any otherwise trivial reason not to live might influence behavior.

Under such unfavorable circumstances, Denys postulated, our best hope for our genetic survival can be, ironically, our death.

*

Before we go any further, it's essential, if for no other reason than an ethical one, that we take a moment to reflect. Whenever I discuss this delicate subject—about the possible adaptiveness of suicide—I do so with great caution. There are devastating possibilities here if people misunderstand the terms being used, relay them poorly, or fail to place them in their proper academic context.

"Did you ever have any difficult conversations with people who thought you meant something else by the word 'adaptive'?" I asked Denys. "Like, 'Hey, well, I was on the fence about it, but now that I know it's adaptive, I might as well go ahead and kill myself.'"

"Yes, I've certainly hit that issue," he told me. "I was always very wary, realizing the sensitivity that suicide occurs in many families and there will always be people in your audience who have firsthand experiences and are sensitive to the issue, and so I was always very careful to express many qualifications."

This is a man who has learned to choose his words carefully over the years. And for good reason. But, again, when you look at it through an evolutionary lens, suicide is just a numbers game. A long bet played out in deep time. And while our wagers may be made blindly, neither are they entirely unpredictable. At least, that's according to Denys.

When evolutionary theorists use the term "adaptive," it has nothing to do with being good or bad in the colloquial or clinical sense of well-being. Nor does it have any moral bearing whatsoever in the sense of being right or wrong. It's just a mechanical term. That's it. Having a fever is royally unpleasant, but it's adaptive under certain conditions; in mammals with raging infections, an elevated body temperature can help reduce bacterial growth and jump-start physiological defense mechanisms. Vomiting from nausea is another miserable but functional case in point. And there are a whole host of unpleasant emotions, such as anxiety and, yes, even shame, that most of us don't particularly enjoy but nonetheless are adaptive in their own right, in that they evolved to motivate our ancestors to avoid certain social behaviors that would curtail their reproductive success.

Although the subject is not without controversy in its own right, some scholars have even argued that depression is an adaptation.

"Constant anhedonia is a hallmark of depression and may reflect the importance of resisting hedonic distractions," argue the evolutionary psychologists Paul Watson and Paul Andrews. The word "anhedonia" may have the ring of a fragrant exotic flower from some far-flung Mediterranean island, but in fact it smells like nothing at all—it's basically that terrible absence of pleasure that we depressives know all too well. So what Watson and Andrews are saying is that this anesthetizing effect on our emotions enables us to concentrate on solving some critically important problem without getting sidetracked by far more pleasant, but unnecessary, happenings. A good night's sleep, a healthy appetite, and a libido are temporarily expendable luxuries that the depressed cannot afford.* I suppose one

*A notable watering down of these primal drives—sleep, appetite, and sex—also means that depressed individuals are not actively involved in large social groupings where they are most vulnerable. Because depression frequently arises after a threat to one's social standing, by being awake at periods when others are not, having a re-

can take solace in considering that being bedridden and worrying nonstop about whatever it is that's currently making one so despondent may be a purposeful, even healthy, response. "Psychomotor retardation" (low serotonin slows physical activity) combined with unrelenting rumination was nature's way to get you to focus on solving an imminent, complex social problem that's seriously threatening your genetic interests.

The social aspect is the key here. Depression, argue these authors, is triggered by problems that are inherently social in nature, or at least those that have big social implications. In line with this, there's a wealth of evidence showing that people in a depressed state are in a hypervigilant social mode, and furthermore this increase in attention to all things social comes at the expense of their basic cognitive functioning.

Generally speaking, depressed people stink at abstract intelligence tests, memory tasks, and basic reading comprehension, but if you're looking for someone who suffers no delusions about this fact, along with everything else they can't do well, they've got it going on. For them, there's a thin line between realism and pessimism, and because they're presently caught in a high-stakes social problem where being overly optimistic about their assets and liabilities can lead to a disastrously bad outcome, it pays, for now, to err on the side of caution and be a Debbie Downer. On those rare occasions when so much is riding on our ability to compute the full range of possible social responses to the major life decision we're contemplating (Should we leave our loveless marriage or not? Quit our stable yet unfulfilling job for a risky but exciting new one? Come out of the closet, whatever type of closet that may be?), and where factoring in worst-case scenarios is a critical part of this reconnaissance mission, being a pessimist isn't without logic.

This also explains why we even have a cognitive *preference* for negative over positive feedback when we're depressed, identifying signals

---

duced need to compete for food and sex, and gravitating to peripheral social spaces (such as one's couch) where physical contact with others is minimal, one can avoid new social conflicts while the present problems are being worked out.

of rejection (facial expressions or frosty emails) more quickly when presented with mixed social cues and mulling over them longer.

In the end, either all of the quiet effort your brain puts into depression leads to a lightbulb let's-do-it realization about what's needed to solve the problem, and you're spurred to take action and get back to the world of the living, or, despite all the mazelike wanderings through countless trails of hypotheticals, tunnels of what-ifs, and dark valleys of worst-case scenarios, you merely arrive at the jarring fact of the matter that there simply is no way out.

So, hopelessness, according to this model, isn't necessarily a bad thing. "Depression should abate when a problem is perceived to be truly unsolvable," claim Watson and Andrews. In other words, at some point, your brain recognizes that you just need to get on with living your life, despite the sad situation. "It is what it is," as that tired old trope goes.

I don't know about that, personally. I endured years of depression without it abating and was helped immensely by taking prescribed antidepressants. The problems underlying my depression didn't go away, but I was able to cope better with them when medicated. Waiting out major depression is a little more serious than the decision to forgo taking two Tylenol and instead just allowing wise old nature to take its feverish course. Still, the logic's the same. "Even when a therapist can implement a helpful talking therapy," Watson and Andrews conclude, "it may be best to let depression work its miserable yet potentially adaptive magic on the social network. . . . [We suggest] that drugs should not be given unless the causative social problems also are being addressed, and that drugs not be allowed to emasculate the ruminative and motivational functions of a potentially adaptive depression."

Well, okay, depression's one thing. But suicide as adaptive? That just seems like it's on a whole different level. With depression, we might even come out with a new, healthier outlook on life if it all works out as nature intended. As Nietzsche said, what doesn't kill us only makes us stronger. But suicide does kill us.

Watson and Andrews concede that a subset of depressed people, around 5 percent, do end up taking their own lives rather than returning to the world of the living. The authors attribute many of these deaths to instances of parasuicidality. If you recall from the previous chapter, these are those cry-for-help cases in which the depressed person doesn't really want to die, but they've underestimated the lethality of their efforts in signaling a need for social support and, wouldn't you know it, now they've gone and actually killed themselves.

Suicide attempts outnumber actual suicides by about 10 to 1. And parasuicides, say Watson and Andrews, can happen if "extortionary depression" (which basically means that, as a partner, friend, or family member, you get to a point where it's better to simply give in to what the depressive wants than continue enduring the costs being imposed on you) doesn't work. When you look at it this way, a serious suicide attempt is what's known as an honest signal in evolutionary lingo. Threatening suicide can be a manipulative tactic, and the reason for countless relationships languishing long past their expiration date.* But attempting it ups the inclusive fitness ante. A suicide attempt literally says: "See? See how dead serious I am?" Watson and Andrews explain:

> As an honest signal . . . the risk of death associated with a suicide attempt could inform partners about the attempter's level of need. . . . Suicide attempts impose a risk of loss on all partners with an interest in the attempter's existence. [The individual] threatens repeated, escalating attempts, exposing partners to further risk unless they provide sought for help. . . . Thus the attempter trades off the risk

*The use of suicide threats is a surprisingly understudied topic. It isn't uncommon for young children to shout in protest, "I'll kill myself!" when some especially undesirable circumstance appears to them impossible to change. (My eleven-year-old nephew shouted these very words when his mother told him that they'd have to get rid of the family dog after it had become unpredictable and bit a neighbor boy; his suicide threat seemed to shock him when it came out of his mouth just as much as it did us horrified adult listeners.) Any suicide threat should be taken seriously. Suicide does happen in preteens, but it's relatively rare. One line of inquiry in this vein might be to determine if having a history of making such threats in childhood predicts suicidality in later adolescence or adulthood.

of death against the possible benefits to be gained from motivating close social partners to help.*

One of the best predictors of suicide, in fact, is a previous attempt. Moreover, many suicidal acts, across cultures, indeed appear to offer support for what the anthropologist Kristen Syme calls the "social bargaining hypothesis," in which suicide attempts are used as a powerful leverage to change the stubborn positions of people who have a vested genetic interest in the actors' well-being. A young woman who tries to drown herself because her family is forcing her to marry a man she doesn't love would fit this bill. She might die in the process, but that's the point: because she's imperiling everyone's genes, it's a strong bargaining tactic likely to effect change when nothing else will.

Here's what a previously suicidal subject told my student Bonnie Scarth about his state of mind during one of his attempts:

> I was alone at home and, you know, I hadn't been going to school and at one point I was in the bathroom and it just became completely unbearable how bad I felt, and I went to the medicine cabinet and got out a big huge bottle of aspirin and poured half of it into a glass and drank it down, and then called for help. I think, what I have come to realize, that of the ten times I did something to hurt myself, there were probably three or four that were really serious and dangerous, and the rest, they were serious, but they weren't actually attempts to harm myself; they were attempts to change intolerable situations.

These types of behaviors are obviously desperate and highly dangerous, but Syme has argued that the emotional strong-arming of others through self-destructive acts can work to the attempter's advantage. "Rather than altering victims' neuropsychology, psychol-

*Seemingly relevant to this is the fact that 78 percent of attempt survivors say they regret the attempt in the days immediately following the incident. Of course, that sort of response could be interpreted in any number of ways (e.g., not completing the act; genuine remorse; deceiving hospital staff about ongoing suicidal intentions, and so on).

ogy, or behavior," she wrote, "the most effective response to suicidal behavior would be to substantially improve victims' circumstances, which in many cases would involve changing the attitudes and behavior of their social partners." In other words, the most immediate way to keep a person from killing themselves, according to Syme, is to bend the will of those who are posing an obstacle to the suicidal person's goals, not modifying the latter's outlook, perceptions, or coping strategies.

Regardless of what you think of such evolutionary models, what's important to stress is that even if the individual is employing these "manipulative" tactics to effect change, it does not mean that they are consciously strategizing to get their way. (Again, unless they're a psychopath.) Their pain is genuine. "When I am making those kinds of decisions," said one interviewee who has often felt this way, "it is not like I am slowly considering, 'If I do this, this will happen. If I do this, then this will happen and here will be the consequences.'"

> It is just, "This has got to stop," and the way I know it is going to stop is by doing something to make it stop. At a certain point, it doesn't matter if it stops for two minutes or two weeks. I just have to have relief from how terrible I feel, you know?*

An analogy is the adaptive "manipulation" of maternal caretaking responses by infants. If you've ever heard a baby really screaming his head off, it sounds like he's experiencing genuine respiratory distress. That's no coincidence, say evolutionary psychologists. Tens of thousands of years ago, ancestral babies who cried like that had an advantage over those who didn't, because ancestral mothers, who, needless to say, had a vested genetic interest in their babies' survival, were more likely to respond to these choking-like sounds. Just like suicidal people threatening or attempting to kill themselves, the emotions driving the behavior—the infants' distress—are real.

"The prototypical suicidal state," wrote Shneidman, "is one in which an individual cuts his throat and cries for help at the same

*Interview excerpt from Bonnie Scarth's unpublished doctoral dissertation.

time, and is genuine on both sides of the act."* The ambivalence of wanting to both die and be rescued only intensifies the suicidal person's despair. The Swedish scholar Kristian Petrov once wrote of a woman who'd been treated for years for her suicide attempts; in the course of her psychiatric treatment, the woman was diagnosed with an unrelated heart condition and got fitted with a pacemaker, and it's then that she began suffering from death anxiety.

There's a logic to such contradictions. One of the great paradoxes, it seems to me, is that nobody wants to live more than the suicidal person, just not under *these* circumstances; no one has a greater appreciation for life, just not *this* life. When suicidal, I'd happily swap places with anyone, or anything, as long as it isn't me. I recall kicking over a log in a sunny field one day and being jealous of the beetles rushing back to darkness. Remove the suicidal person's one nagging reason for dread and the desire to die evaporates . . . for the time being.

The trouble is that the types of problems leading people to take their own lives tend to be complicated. It may be sound advice, but in the majority of cases, Syme's suggestion to change the social en-

*The most common assumption about the well-known sex difference in the lethality of suicide methods is that women just aren't as serious about killing themselves as men—that it's instead a "cry for help" or they're "just doing it for attention." They don't *really* want to die, goes this line of reasoning. It seems rather odd to be arguing for gender equality when it comes to being really and truly suicidal. But some researchers point out that suicidal women really do want to die, at least as evidenced by recent female attempters' scores on "The Suicide Intent Scale," a checklist including such items as "did not contact or notify helper before suicide attempt" and "took active precautions against discovery, such as a locked door." No, it's not intent that's the issue, these scholars claim, but that women's suicidal missions are more likely than their male peers to be compromised by vanity, namely, their concerns about their postmortem visage. If one believes (wrongly, I hasten to add) that an overdose creates a supine pose like that of Sleeping Beauty, gorier manners of death may be passed over for these superficial reasons. A less sexist-sounding twist on these data is that suicidal women don't want to leave their mutilated bodies behind, anticipating the trauma it will cause for whoever finds them. I don't know if "considerate" is the word, but whatever it is, more men who commit suicide fail to possess the quality than women.

vironment, rather than the person's fit within it, is easier said than done.

✷

What about the many people who aren't using suicidal behavior as a desperate bluff, but clearly—by employing highly lethal methods in places and at times where no one can possibly save them—have every intention of dying?

Denys's model acknowledges that many suicides are unintentional (or at least, not quite intentional). But in contrast to Syme's social bargaining hypothesis, he argues that even those that clearly aren't a cry for help still meet the definition of an adaptive response. Perhaps the clearest example of how his model works in real life is "altruistic suicide" among the elderly, a behavior documented by numerous ethnographers around the globe. About sixty years ago, one of these ethnographers, the Danish anthropologist Knud Rasmussen, asked a Netsilik Inuit from the Canadian far north why so many old clan members chose to die by their own hands. The respondent explained patiently: "For our custom up here is that all old people who can do no more, and whom death will not take, help death to take them. . . . [T]hey do this not merely to be rid of a life that is no longer a pleasure, but also to relieve their nearest relations of the trouble they give them."

It's not that such cultures don't value their elderly. On the contrary, until the person's decision to hasten their own ending, all lives are highly regarded; even then, those who die selflessly this way are expected to be rewarded in the afterlife, similar to martyrs or children. In regions where resources are relatively plentiful and elderly family members can still compensate in meaningful ways, such as helping to care for their grandchildren or instilling cultural knowledge, suicide rates among the aged are lower than they are in other nations.

A similar interpretation has been used to explain the riddle of suicide attacks in war-torn regions. When it comes to military strategies, evolutionary theorists working in this area, such as John Orbell and Tomonori Morikawa, obviously factor in different variables from those applied to altruistic suicides of the elderly, but the same strong undertow of kin selection is believed to be at work. After carefully coding hundreds of diaries and letters of Japanese Kamikaze fighter pilots in the closing months of World War II, these authors concluded that what motivated these otherwise sane pilots to, so to speak, fly in the face of traditional Darwinian thinking and guarantee their own oblivion by dive-bombing enemy fleets was the powerful notion that such a drastic maneuver was the last option remaining to save their family members back at home. In every major battle, explained Orbell and Morikawa, a warring faction is a conglomerate of kin-based coalitions that implicitly agrees to work as a collective serving the individual genetic interests of each soldier (emotionally fueled kin-based rhetoric, such as "band of brothers," effectively exploits this psychology). After all, the obliteration of one side inevitably meant a terminal end to each combatant's fitness. And in the ancestral past, the near wiping out of an entire population in a wartime conflict—what we call genocide today—probably wasn't all that uncommon.*

Still, none of this tells us anything about how, for lack of a better word, everyday suicides—those impulsive, rash self-endings driven by a violent surge of negative emotions, or even those meticulously planned-out ones by those who've thought long and hard about it—can be explained in such adaptive terms.

Suicides, that is, like Gregory's.

In the 1990s, Denys conducted a series of studies testing his hypothesis that suicidal thinking is most common in people facing poor reproductive prospects and simultaneously consuming resources without contributing to their family. For the most part, his predictions were accurate. Participants who were currently suicidal were also more likely to report feeling they were a burden to their

*Central to their assertion, however, is that this type of last-ditch military decision occurs only when the losing party (in this case, the Japanese) has some hope of not being wiped out entirely in a genocide. If genocide is seen as inevitable, "participation in a suicide attack would make no adaptive sense at all," the authors argue. "Adaptive sense would appear to reside, instead, either in saving significant numbers of one's close kin or, if that were not possible, in saving oneself."

families, to have had fewer children, less heterosexual sex in the past month, fewer friends, and more financial problems. For women, having children has long been a well-known protective buffer against suicide, but more recent findings indicate that this applies only to mothers whose kids still live at home with them.*

"Did you ever learn why Gregory took his own life, what he was going through at the time?" I asked Denys, curious to find out if his brother's death fits his own model.

"Well, he didn't leave a note," Denys told me. "But there were certainly a lot of things going on in his life at the time. Heavy things."

The previous summer, Gregory had been staying by himself at the DeCatanzaro home in Evanston for a few weeks while the rest of the family was away on a trip. During that time, he'd apparently become sexually involved with a neighborhood girl, possibly underage. She was also Roman Catholic and didn't believe in birth control. "We don't know for sure, but we suspect she might have gotten pregnant

*There are plenty of anecdotes but, rather surprisingly, only one study that has directly investigated the connection between pet ownership and suicide rates. That 1985 study in the *American Journal of Public Health* looked at Maryland residents who'd died by suicide between the years 1975 and 1983, and found they were no more or less likely to have owned a dog or cat than were living control participants. "Whatever may be the positive health effects of pets, they seemed not large enough to affect suicide substantially in this population," write authors Knud Helsing and Mary Monk. There were significant limitations to this old study, however. For one, the researchers did not measure these people's attachments to their pets. They relied entirely on informal county census measures indicating simply that a dog or cat was or wasn't living in the home in the year 1975 (it could have belonged to another family member). More importantly, the presence of a dog or cat in the home in 1975 doesn't mean that the same animal (or any other animal) was there in the seven subsequent years for which the relevant suicide stats were considered. For all we know, then, a person's suicide in January 1976 may have been precipitated by the loss of their cherished pet in December 1975. One would think, after all, that since pets are basically proxy kids for so many people, the protective function would tend to operate similarly. One woman had this to say about how her pet rabbits gave her the will to go on living: "They are something that is very important in my recovery and helping me not get too depressed. Even when I was so depressed, I was kind of suicidal. I never got really bad, but I was suicidal at one time. The thing that made me stop was wondering what the rabbits would do. That was the first thing I thought of and I thought, oh yeah, I can't leave because the rabbits need me. So they were playing a really big role in that."

and given the baby up for adoption," Denys told me. "My parents of course never knew any of this until after the fact."

As if that weren't enough drama for a high-strung nineteen-year-old, when he returned to school in Halifax, Gregory had started dating a graduate student. "He was head-over-heels in love with this other woman," said Denys. "She was beautiful." But things got complicated when she revealed to him that she was afflicted with a congenital disorder, one that caused her to have ambiguous genitalia. "On discovering this," Denys told me, "Gregory—being a teenage boy—experienced a real shock in intimacy."

With such weighty problems occupying his mind, Gregory found himself distracted from his schoolwork, and he nearly failed his exams in the spring of 1965. His poor academic performance only compounded his anxieties. "I think he would have felt shame and apprehension at the thought of having to return to family and Evanston under those conditions," Denys said. "And there's probably more we don't even know. Actually, a week or so after his death, my mom received an odd local phone call from a young woman who rather aggressively asked to speak to her eldest son. We never found out what that was about. So yeah, there were definitely some major things going on in his life."

"And back then—" I said, remembering that we're talking about 1965.

"Right," replied Denys. "Social mores were much more conservative then, his father was a religious leader in the community, and his problems may have seemed insurmountable to him. And there was something else, too," Denys seemed to suddenly recall. "Everyone in the family had obtained green cards in 1959 when we first moved from Canada to Illinois. That included Gregory, even though he didn't accompany us at that time. So, he was required by U.S. law to register for the military draft when he turned sixteen, but he never did. After failing his exams, he was supposed to come home to be with us, but he may have also feared crossing the border at that time because of the possibility of being arrested for draft evasion. Remember that the U.S. involvement in the Vietnam War was intensifying in 1965."

"What a mess," I said.

"Yeah," said Denys, rubbing his eyes in resignation. "I guess I'll never really know what was going on in my brother's head."

*

Not all scholars have reacted favorably to Denys's ideas about suicide, or to any theory, for that matter, that sees suicide as an evolutionarily adaptive response. The suicidologist Thomas Joiner is a case in point. In a recent critical review, Joiner, an influential figure in the field, concludes that suicide is "an exemplar of psychopathology," which is to say, a case of nature gone awry and unraveling at its seams, not one unfolding according to adaptive design. In doing so, Joiner is advancing the more traditional disease model of suicide: it is a sickness, not an adaptation.* Joiner and his coauthors arrive at this conclusion by dint of the painfully true fact that suicide is so ruinous for those left behind. Adaptive? How can suicide be adaptive, they ask, if the reproductive potential of bereaved relatives, who are more often than not left permanently hobbled by traumatic grief, is inherently worsened by a suicide, not improved?

At face value, it's certainly hard to disagree with Joiner's camp. The authors state that *everyone* ("100%") who dies by suicide has a mental disorder. They then list the many reasons why they've come to favor this rather stark statistic. They observe:

> Suicide involves the unsanctioned and frequently brutal killing of an innocent; the potential deaths of others via suicide contagion, not to mention the occasional actual deaths of bystanders (e.g., those landed upon by . . . jumping from a height in an urban setting, those killed by chemical exposure); the deprivation of choice and future care and comfort to loves ones; and the willingness to devastate dozens of people into a shocked state of bereavement, not infrequently without warning and certainly without their consent.

*For a fascinating historical review on the medicalization of suicide, see Ian Marsh, "The Uses of History in the Unmaking of Modern Suicide."

"Any one of these is suggestive of psychopathology," summarize the authors. "Their conjunction is a clear exemplar of psychopathological functioning."

Well, perhaps. On the other hand, there are plenty of behaviors that are bad for the collective whole—ugly, horrible aspects of contemporary society even—but they're still probably grounded in adaptive responses to ancestral conditions. For example, infants show a clear preference, a prejudicial bias, for people of their own culture. When given the choice between two strange but smiling adults who are offering them a fun new toy, *preverbal* infants will spurn the one speaking to them with an unfamiliar accent, ignoring eye contact with this "foreigner" and refusing to reach out and accept their gift. Developmental psychologists have suggested that this innate preference for native speakers, and an inborn distrust of those with accents, makes evolutionary sense. Because the phonetic aspects of languages evolve at a lightning pace compared to biological traits, accents and dialects, which for most of us are quite hard to fake, would have been a rapid-fire heuristic for knowing a hostile outsider from a friendly in-group member tens of thousands of years ago in small-scale societies rife with intergroup conflict. The fact that such social biases today contribute to selfish, malignant ideologies of xenophobia and racism in complex multicultural cities doesn't mean that those who don't override these biases have a mental illness. It just makes them bigots with the unreflective brains of babies.

So to suggest that people who die by suicide have a psychopathology because, ipso facto, dying this way harms others would seem to insert moral reasoning into the adaptiveness equation. And, again, that's a place where it simply doesn't belong. Evolution is an amoral, mindless machine. Whatever works, works.

Even so, does suicide really *work*? Denys's theory has some pretty big holes, actually, especially if suicide impairs genetic relatives' reproductive success through, say, stigma. For example, in the 1960s, the psychologist Richard Kalish administered a "social distance scale" to measure college students' prejudicial attitudes toward a real hodgepodge of stigmatized communities. One of the questions in the scale was "Would you willingly go out on a date with [this type of person]?" Participants were more willing to date someone dying from cancer and members of a marginalized ethnic or religious group (in this old study, Blacks, Mexicans, and Jews) than they were someone who'd attempted suicide. On the other hand—and I'm not sure if this qualifies as good news, exactly—they were more willing to go out on a date with a suicide attempter than a Nazi. And when Kalish's study was replicated twenty-five years later by the suicidologist David Lester, the trends were identical. Furthermore, when asked, "If you really loved him/her, would you marry someone who had attempted suicide in the past year," only 33 percent of people said yes.

If this type of prejudice and social distancing toward suicide attempters extends to the close family members of those who actually die by suicide, as is often claimed by bereaved survivors, this could pose a significant problem for Denys's adaptationist logic. One woman who'd lost her father to suicide six years prior wondered aloud, "Will the stigma be attached to the children, to the children's children, and to their children in turn?"

Still, studies done on college campuses don't necessarily reflect default human attitudes, and at times and places where suicide is (or was) less taboo than it is today in much of the world—such as, say, in ancient Greece, where the topic was discussed in a vein of icy reason—those stigma data as reported by Kalish and Lester may have looked very different.

In any event, let's place Joiner's "suicide is decidedly not an adaptation" argument in proper context. To begin with, he sees human beings as a *eusocial* species. In many ways—in most ways—we're like overgrown ants. Eusocial insects, such as ants, termites, and many types of bees and wasps, live in busy hives complete with multigenerational social caste systems. Typically, there's royalty, such as a queen, along with high-status breeders and a legion of sterile workers united through a division of labor, such as soldiers, hunters, and builders. A defining hallmark of eusociality is cooperative care of the young.

By a long list of criteria, human social systems are remarkably ant-like.* And Joiner, like many other scholars, in fact, believes that the closest analogy to human suicide can be found not in the anthropomorphized deaths of beasts of burden, nineteenth-century pets, or orphaned chimps, but in the fatally self-destructive behaviors of eusocial insects. "[There are no] eusocial species that lack a behavioral repertoire for highly lethal self-sacrifice under conditions of inclusive fitness," writes Joiner. An individual honeybee, for example, will give its life by stinging a threatening target encroaching on the hive. Such a bee always dies, usually within hours. After all, in depositing its stinger in a bear's nose (or in the thumb of my six-year-old self on the school playground back in 1981), it also loses almost half its body. If the pain of one bee's sting alone is insufficient to keep the interloper from molesting the hive, which is a repository of the individual honeybee's genes, the pheromones released by the stinging event act as a rallying cry for other bees to attack.

In other cases, meanwhile, an individual organism infected by a parasite will act for the good of the group to keep their non-parasitized kin from getting infected. The beefier bumblebee, for instance, can sometimes get contaminated by a nasty little parasite called a conopid fly, which inserts larvae into the unwitting bumblebee's abdomen. This rude act kills the infected bumblebee within twelve days, and the larvae pupate in its desiccated corpse until they hatch. The zoologist Robert Poulin, an expert on the evolutionary arms race between parasites and their hosts, has found that when a bumblebee gets infected by one of these murderous buggers, it often does something remarkable: It abandons the colony and flies off to a faraway flower meadow to die completely alone. Why? Because in doing so, the invidious larvae are also led away from the infected bumblebee's non-parasitized kin, and the colony is saved from an infestation when the conopid flies hatch.†

*The authors suggest that human females past reproductive age are roughly analogous to sexless worker ants, which is, well, an interesting way to look at menopause, if nothing else.

†Poulin cautions against use of the term "suicide" when describing the self-disadvantageous behaviors of other species, particularly insects: "The adoption of a more dangerous lifestyle by an insect that is bound to die shortly may be adaptive

Joiner's claim that suicide is the product of a diseased mind, rather than an adapted mind, reflects his view that humans are eusocial. Because one person's suicide is so bad for the rest of the hive, so to speak, by triggering possible copycats (more on that in chapter 6) and causing so much grief and misery, suicide isn't an adaptation but instead a "derangement of the eusocial instinct." Suicidal people, he reasons, are acting *as if* they are infected ("social toxins") and remove themselves from the group after falsely concluding that their deaths are more valuable than their lives. "These individuals," write Joiner and his colleagues, "act in response to a misperception of Hamilton's rule, believing that their own lives must be sacrificed in order to benefit society as a whole. This misperception is itself the derangement that is at the core of our argument [that suicide is not an adaptation]."

There's much that I like about Joiner's theory. The eusocial framework helps us to appreciate why it's so important to convey to suicidal people that they have irreplaceable qualities that can benefit others. They have a role, a place. They are needed in society, even if their function feels peripheral or they've lost sight of it altogether. It's probably not wise to liken them to an ant with a purpose at this fragile emotional stage, but you get the idea.

The trouble with Joiner's argument, however, is twofold. First, it's just kind of a nonstarter, because most eusocial insect species, unlike human beings, are *haplodiploid* organisms. Hives are filled with identical, or near-identical, twins.* Since eliminating yourself as a bee

---

in terms of inclusive fitness, but no more suicidal than, for instance, an ageing animal taking risks to reproduce in the presence of a predator as its inevitable death approaches."

*Because the females in these species develop from fertilized eggs and males from unfertilized eggs, a male's daughters possess all of his genes and half their mother's. Sisters therefore share 75 percent of their genes with each other, which is why Hamilton liked to call them "supersisters." So if you're a haplodiploid female, being a sterile worker and helping to raise your sisters is going to be a much more direct way to pass on your genes than actually bearing offspring yourself, each of which would have, by contrast, only half your genes. Thus, for all the behavioral similarities between us and eusocial insects, animals for which only a select few males and but a single female—the queen—out of an entire buzzing, blooming population will ever reproduce, there are also unequal evolutionary dynamics. This is especially relevant when it comes to self-sacrificial behaviors in eusocial insects.

or an ant in the hive is not really "eliminating yourself"—after all, there are tens of thousands of genetic copies of you in reserve—the selective pressures operating on these instinctive eusocial behaviors simply aren't analogous to those that (may) underlie human suicide. They're both evolutionary long bets, but they come with very different odds.

The second problem I see with Joiner's model is that it presupposes all suicides are the result of misperceptions of Hamilton's rule. That strikes me as idealistic, unfortunately. Despite public niceties, if we stick to using genetic currency and define terms such as "burden" and "worth" using purely mathematical means, it's likely that some people really are a burden to others, and their deaths are worth more than their lives. Does that mean that they *should* kill themselves? It means no such thing. Evolutionary theory isn't a deontology.

Furthermore, if misperceptions do frequently arise, with some or even many suicides being the result of people overestimating their burdensomeness, it's not necessarily a derangement of their evolved eusocial drive. Rather, such misperceptions could also be the result of a disconnect between the social environment of the Pleistocene (the geological period spanning 2,500–11,700 years ago) and that of today. It's easy to be distracted by all the shiny technology we get to hang our cognitive hats on, but under our skulls, we're still operating with those basic brains initially suited for our hunter-gatherer ancestors. These ancestors of ours dwelled in groups of just a few hundred individuals. They were people who lived and died knowing each other—and, more importantly, talking about each other—their whole lives long.* When led to believe that you're intrinsically

*Now imagine living under these same limiting ancestral conditions and being told every day, by nearly everyone, that you're not wanted. That you are bad. Ugly. Cursed. Evil. That you disgust the other group members. Perhaps it's something you've done. Or maybe it's just something they think you've done. Sure, intellectually, one can step back and take some broader perspective about one's still valuable role in some abstract "society," despite what everyone is telling you, but that would have been an unlikely coping mechanism for our ostracized ancestors, one requiring philosophical insights difficult to acquire even today. Likewise, it's a tall order for a fifteen-year-old being bullied relentlessly over social media. Referring to his suicidal response to this modern-day scenario as psychopathological because of his "willingness to devastate

flawed and poisonous to the group, this may signal that your family members' reputation is similarly marred by association and that your suicide, perhaps by freeing them from a relationship with you, can redeem them. Perhaps, to your ancestral brain, it will mitigate retaliation against them for whatever it is you've done wrong, indicating to the group that they've suffered enough by your loss. In many cultures, there's a tendency to forgive the dead; hence that Italian mortuary aphorism *de mortuis nihil nisi bonum* (Of the dead, [say] nothing but good).

The anthropologist Bronisław Malinowski once described how in some small island communities, those accused of violating a serious cultural taboo would climb to the top of a palm tree, declare their hurt at having been charged with such an offense, and plunge headfirst to their deaths, a form of institutional suicide often salvaging their relatives' standing within the group.* Many "shame-based" Asian societies have similar long-held traditions, such as *seppuku* among the ancient Japanese samurai.

Even when cultures lack such explicit expectations of suicide as a way to lessen the reputational damage of familial disgrace on one's surviving kin, this doesn't mean that the same inclusive fitness mechanisms aren't at play in motivating suicidal thoughts. Under such conditions, some people *are* social toxins, not simply responding *as if* they are. In his book *Stigma: Notes on the Management of a Spoiled Identity*, the sociologist Erving Goffman wrote that simply being seen in the company of a social pariah is like "having smallpox." Stigma contaminates everyone in the offender's close circle and, as such, may have real consequences for these people's reproductive success when considered in evolutionary terms. (If you doubt this, go grab a coffee at your local Starbucks with Harvey Weinstein and

---

others into a shocked state," or his failure to realize it may cause others to do the same, feels off. It seems to me, rather, his evolved brain is leading him to compute a Hamilton's value essentially verbatim to what he's hearing.

*I realize this may appear to contradict what I suggested earlier, about family members of suicide victims being stigmatized. But it's easy to see how such suicides could ameliorate an even more damaging stigmatizing effect caused by the kin's serious wrongdoing. In other words, it would be the lesser of two stigmas.

Bill Cosby, and let me know how that goes.) When the offender is a family member, the potential damage is even more severe. Say, for instance, that your sibling or parent is publically accused of a serious criminal offense such as rape, child molestation, or murder. It's splashed all over the news. Everyone knows. And everyone is furious. If your criminal relative now has suicidal thoughts, it could be that they just wish to escape punishment; but their suicidal thoughts could also be seen as motivating an evolved behavior in which "self-extirpation" from one's kin group minimized the contaminative effect of their transgression on their inclusive fitness.

That being said, I suspect that the mismatch between our ancestral and prevailing conditions has led to people overestimating the relative seriousness of their social offenses and problems. That shibboleth "suicide is a permanent solution to a temporary problem" has always rubbed me the wrong way (after all, some problems are quite permanent, thank you very much; it's your perception of the problem that matters), but it rings true for, say, teens dealing with some intolerable, but passing, drama with their classmates.

Contemporary high schools are highly artificial social environments that clash with our evolved brains. As such, they're just asking for trouble. If we dare to corral hundreds of adolescents into long-term facilities in which nearly all of their meaningful relationships are expected to be with others born within twelve narrow calendar months—converging in space at the exact stage in development when sexual competition begins to peak (and along with it, displays of verbal and physical aggression)—we shouldn't be too surprised to learn that bullying is more or less an intractable problem. For a lonely freshman convinced that she's unlikable, unattractive, and worthless because that's what she's been hearing from her hormonally addled peers in the hallways for the past year, such negative feedback pushes those ancient evolutionary buttons in her highly social brain, creating the false impression that she's a pariah or has done something unforgivable. This everyday feedback may lead her to generate beliefs, including suicidal thoughts, out of proportion to what we all come to realize in the end: *In the scheme of things, high school doesn't really matter.*

Nevertheless, for her, at this time, in this place, such a contrived social system certainly feels like "the real world." And she's not alone. In 2011, 16 percent of American high school students said that they'd seriously thought about attempting suicide, and 8 percent actually did try to end their lives. (We'll talk more about the role of social media in a later chapter, but "cyberbullicide"—bullying through online social media that's linked directly or indirectly to the target taking their own life—is yet another evolutionarily unprecedented social problem affecting the modern teenager.)

In other cases, suicidal ideation arises from an all-too-accurate perception of one's standing in the current social climate. Consider an email exchange I had with a reader who'd stumbled across my *Scientific American* article on suicide from a few years back. "I'm 49 years old," began this writer, whom we'll call "Mike." "I did seven years in prison—no crimes before or after. The [sex offender] registry has been killing me softly with its song":

> ... I have been living in a barn for the last 13 years, taking care of this property, and glad for it. But my landlord recently died and now I'm faced with going out into that world. I cannot come up with an argument that is better than suicide. I have come to view people as threats and I cannot imagine a way to live without the thought of impending doom.

This self-exiled man hasn't misperceived anything. He may be mentally ill for other reasons—there's not enough to go by here, and I wasn't about to push for the sordid details of his offense—but it's certainly not because he's contemplating removing himself permanently from a society that has done everything it can to inform him that he's basically plutonium. No matter what good he does from this point on, his crimes will overshadow it. By becoming a recluse, and presumably distancing himself from whatever family he has left, he's already committed metaphorical suicide, and now he can no longer even do that.

In fact, the circumstances always vary, but I've received many such emails from desperate readers, readers who really *have* gotten themselves into very tricky situations where, frankly, it's hard not to see suicide as a rational decision. What's important to point out to these people is that there's no imperative to be rational; if it means saving lives, I'm more of a pragmatist. Even if you favor an evolutionary outlook, there's no rule saying that we *should* or *ought to* act adaptively. Evolutionary psychology is the language of math, not morality.

Still, what does one say to an individual such as Mike? What would you say? To tell him that it's all in his head, that society doesn't really hate him as much as he thinks it does, would be patronizing and dishonest. (Just read the comments section on any news story about a sex offender.) I could only advise him to seek out a clinician without a retributive mind-set and to try to find a pro-social purpose.

"It sounds trite to say 'life isn't easy,' " I wrote back,

> but, nonetheless, it's true, and for some it can be unbearable. If I can impress anything upon you right now that makes you reevaluate your options, it's this: Despite your present feelings of ostracism and isolation, there are more people than you may know who are in your corner. Sometimes the world can seem like it is filled with enemies, but when you lay yourself out there completely, allowing yourself to be honest and vulnerable, as you've done in your email to me, you will find people who will surprise you with their kindness and compassion. . . . You're responding emotionally as any person would under a similarly challenging set of circumstances. You can still come out stronger for this, and maybe help others down the road.

Alas, Mike saw the situation with grim clarity. "I appreciate what you've said," he wrote,

> but I am very smart and can fill in the blanks of what happens after I do what you suggest. The only workable solution I can imagine is that somewhere someone has a large enough property, and enough work they needed doing that I could live there away from anyone's worry of being near a sex offender, in exchange for labor. I'm a painter and an electrical ace with all the tools to do what's needed; I'm artistic and do my work with integrity. I don't imagine myself being anyone's

friend. . . . I have lived a life of desperation from childhood, but I try to trade value for value.

I'd be lying if I said the idea of a kitchen renovation didn't pass briefly through my head; I'm all thumbs and, let's be honest, probably not the guy's type. But what's useful to recognize in this example, and any other case where suicidality is precipitated by social anxiety (which stems from a fear over others' unpredictability), are the emotions on display. They center on *other people's thoughts about the self.* Shame, humiliation, guilt, embarrassment—that same batch of social emotions we explored in the previous chapter.

Worrying about others' worrying about us is painful.

In biology, a distinction is often made between "ultimate" and "proximate" causes. Ultimate causes are those subconscious pressure systems propelling organisms toward adaptive behaviors, whereas proximate causes refer to the immediate, usually emotional, factors that prompt the organism to action. For instance, consider the question "Why do we have sex?" The answer in terms of ultimate causation is that sex can lead to offspring, and therefore it's the best way to ensure our genetic interests. The answer to the same question in terms of proximate causation, by contrast, is that, well, sex feels really good. To be a well-adapted animal, one certainly doesn't need to deliberate about their reproductive potential. What motivates us is the promise of an orgasm, a powerful physiological response of euphoria that serves as the proximate hook for an ultimate gene-replicating purpose.

In ultimate terms, human suicide may or may not be bound to patterns of inclusive fitness, just as it is for the self-destructive behaviors of eusocial insects. Yet for us, such behavior isn't normally triggered by predators or parasites, but more often other members of our own species. The proximate cause, in other words, appears most often to be our desire to escape from the feelings caused by other people. We seek to hurl ourselves into the obliterating abyss in reaction to felt negative judgment. The stuff of other people's minds is the rarified emotional fuel behind this terrible death instinct. If caught in the throes of it, keeping a genuinely suicidal person from completing the act may be as difficult as encouraging someone at the very peak of sexual excitement to please kindly refrain from having an orgasm, which is itself sometimes referred to as *la petite mort* (the little death).

*

Denys, for his part, agrees with Joiner about social integration being vital to suicide prevention efforts. But unlike Joiner, his evolutionary reasoning is based in a more traditional inclusive fitness theory, not in terms of a literal form of human eusociality.

"Well, bottom line," Denys told me, "is if I was asked to give advice to someone whose loved one or whoever is contemplating suicide, I would say give them a sense of worth and purpose. Because I think that that—"

"Alters their perceptions?"

"Yes, and it probably evolved as a proxy for inclusive fitness maximization. I think many people can get a feeling of self-worth out of contributing to the community, and that feeds back into the broader adaptation related to benefiting biological family members. I think work that is appreciated is probably the best guard against suicide for someone who may not have opportunities to reproduce or to contribute to their kin directly."

Denys's advice reminds me of a scene from Jean-Jacques Rousseau's epistolary novel *Julie, or the New Heloise*. In it, an older man, a pious cleric, is counseling a suicidal younger man, a philosopher who sees no good reason to live and wants to put a rational end to his sufferings, just as the Stoics did in ancient times. "Listen to me, mad youth," says the wise old man. "You are dear to me; I pity your errors."

> If you still have deep in your heart the least sentiment of virtue, come, let me teach you to love life. Every time you are tempted to exit it, say to yourself: "Let me do one more good deed before I die." Then go find someone needy to assist, someone unfortunate to console, someone oppressed to defend. . . . If this consideration

holds you back today, it will hold you back again tomorrow, the day after tomorrow, your whole life long.*

It's compellingly pragmatic advice. More than that, however, it's supported by empirical evidence. When we feel needed, we're less likely to take our lives. This may help us to understand the counter-intuitive but well-known finding that suicide rates tend to plummet during wartime, when there's a shift in the cultural focus away from individual differences and toward the unification of in-group members. A similar precipitous drop in America's suicide rates occurred in the immediate aftermath of events such as John F. Kennedy's assassination, the *Challenger* space shuttle explosion, and the 9/11 attacks. This effect of group cohesion has even been used to explain why the suicide rate was lower than one would expect among those imprisoned in concentration camps during the Holocaust.†

"To be rooted," writes the philosopher Simone Weil, "is perhaps the most important and least recognized need of the human soul."

*The next sentence in Rousseau's story goes like this: "If it does not; die, you are nothing but an evil man." But the trouble with that last part, as we'll see in chapter 4, is that the suicidal mind-set is intrinsically egocentric and effectively strips our capacity to care about others. Those who read this and aren't swayed by the old man's logic may mistake their temporary social apathy for "evil" and feel even worse about themselves. It's the type of literary passage that rings true for us when we're not suicidal, but when you're actually in that state, it's not so straightforward.

†The low-suicide rates in Nazi concentration camps have been widely accepted and the subject of considerable scholarship, including by the Jewish thinker and Holocaust survivor Victor Frankl, who attributed prisoners' miraculous capacity for survival under such barbaric conditions to the innate human ability to find purpose in life. The notable suicidologist David Lester attempted to tease apart the mystery of why suicides were so infrequent among Jews during their imprisonment (but, paradoxically, common after they'd regained their freedom) in his 2005 book *Suicide and the Holocaust*. However, a recent review article by Francisco López-Muñoz and Esther Cuerda-Galindo calls into question the low-suicide axiom, arguing that such deaths in Jewish ghettos and concentration camps were in fact quite high but would have gone unreported, been covered up by guards, labeled as failed escape attempts, or disguised as "natural" (e.g., intentional self-starvation).

✷

One of the more surprising early critics of Denys's theorizing about suicide as an adaptation was Richard Dawkins, who, in response to Denys's article back in 1980, countered that suicide could be better understood as a by-product of modern life, one with no clear evolutionary provenance. This he referred to as the "domestic animal hypothesis":

> A domestic animal lives in an environment other than that in which its genes were naturally selected. . . . Darwinians do not worry unduly about moths' habit of immolating themselves in candle flames. Candles, indeed non-celestial light sources generally, are recent innovations in the world of moths, and we are prepared to accept that natural selection due to candles has not yet had time to work on their gene pools. The very sensible orienting technique of maintaining a fixed compass bearing towards light rays coming from infinity can become suicidal if the moth tries to do the same thing to light rays radiating out from a candle. . . . My first impulse whenever somebody brings up some odd aspect of modern human behavior and defies me to "explain that with your selfish genes if you can" is to tell the moth story. That is my inclination in the case of suicide. Darwinians need worry about suicide only if it is under genetic control, and then only if it is widespread in wild animals (humans), observed in the environment in which their genes were naturally selected.

Over three and a half decades later, we now know that suicide is, in fact, under significant genetic control. Looking at twin studies, it's far more likely that identical twins (who share 100 percent of their genes) will both die by suicide than fraternal twins (who share only half their genes), and an adoption study found that adoptees who attempt suicide are six times more likely than non-suicidal adoptees to have biological family members who killed themselves. The latter can't be due to social learning, since in most cases, of course, these genetic relatives are strangers the adoptees have never met.

What's especially fascinating about these genetics studies on suicide is that what gets passed down from one generation to the next isn't just some generic psychiatric condition in which suicide tags along as a corollary outcome, but a predisposition for suicidality specifically. That is, even when you control for things like depression, bipolar disorder, and alcoholism, the tendency to be suicidal still reveals its very own genetic basis, one that can be carefully teased out from these and other heritable individual differences.

What this *doesn't* mean is that suicide is inescapably determined by genes. Even people whose identical twins are lost to suicide are much more likely not to kill themselves than they are to do so. It simply means that when coupled with known risk factors—such as a history of physical or sexual abuse in early childhood, neglect, parental loss, drug or alcohol addiction, and other traumatic experiences over the life course—some individuals apparently have a lower threshold for becoming suicidal than others. It's that sensitivity that's inherited.

These genetics data aren't direct evidence for Denys's evolutionary theory for suicide, but they do satisfy the first of Dawkins's two criteria for taking the theory seriously: it is, to use his phrasing, at least partially "under genetic control."

*

Dawkins's other point, about whether suicide ever occurred in our ancestors ("wild humans"), is more difficult to address. "When did a conscious, deliberate attempt to die veer off from the borderlands of extreme recklessness and impetuous, life-threatening taking of risks?" asks Kay Redfield Jamison in her book *Night Falls Fast*. "Violence and recklessness, profound social withdrawal, and self-mutilation are not unique to our species. But perhaps suicide is. We will never know who or why or how the first to kill himself did (or herself; we will never know that either)." Ancient prehistoric human remains are extremely uncommon finds in archaeology to begin with, and the preservation of unambiguous clues at field sites enabling investigators to give a verdict of suicide is an even more unlikely discovery.

There is, however, anthropological evidence for suicide in almost every society ever studied, including hunter-gatherers, whose lives, and livelihoods, are thought to closely mirror those of our ancient predecessors. Admittedly, colonial-era ethnographies are inherently problematic, both for methodological reasons and racist ones, but in 1894 the Dutch sociologist S. R. Steinmetz published "Suicide among Primitive Peoples" in the *American Anthropologist,* a paper surveying reports of suicide in nearly all small-scale societies for which records were then available. "It is the opinion of many sociologists, who perhaps have not given especial thought or study to the subject," wrote Steinmetz, "that the act of self-destruction is infrequent among savage peoples. The purpose of my inquiry is to determine whether this opinion has the support of well-authenticated facts, and, if so, to what degree."

It didn't, it turned out. Suicide was ubiquitous. From Cherokee Indians ending their lives after being disfigured by smallpox, to pregnant unmarried girls of the Caucasus hanging themselves because the infamy of such a status was so great, to aboriginal men on the Indonesian island of Sulawesi killing themselves upon discovering that their wives were unfaithful, Steinmetz found plenty of material to disconfirm popular opinion.

In fact, he surmised that so-called primitive societies had *more* suicides, a speculation without much merit. "It seems probable from the data that I am able to collect," the author wrote, "that there is a greater propensity to suicide among savage than among civilized peoples, and that its frequency may be owing to the generally more positive faith in the future life existing in the former races which enables them to meet death with greater calmness and a slighter resistance of the instinct and other natural motives tending to conservation of life."

Such an arbitrary dividing line between the "rational" Westerner and the "irrational" savage is more than a little rankling. If those wholesale Bibles on the nightstands of every cheap motel room from Duluth to Dublin are anything to go by, it seems to me that the vast majority of the former are pretty convinced there's life after death, too.

Still, Steinmetz's work shows that suicide has been a part of the human story for as long as, and wherever, records have been kept. Sadly, so many traditional societies were eradicated in the previous century that there aren't many left for anthropologists to study these days, especially hunter-gatherers. But a 2016 study by Kristen Syme, whom we met earlier with her social bargaining hypothesis, confirmed Steinmetz's findings of the cross-cultural prevalence of suicide. Together with her colleagues Zachary Garfield and Edward Hagen, Syme scoured the Human Relations Area Files (HRAF), the most complete archive of historical ethnographics available to researchers, for any and all references to suicide. Instances of suicide emerged in cultures representing all of the major modes of subsistence, including hunter-gatherers, horticulturists, pastoralists, and intensive agriculturalists. Guns, drugs, trains, and other modern methods altered the landscape dramatically and made suicide easier (or at least swifter). Yet there were always plenty of methods available to our ancestors. Jumping, starvation, exposure, drowning, ingesting poisonous plants, hanging, offering oneself to a hungry predator—they're all just different routes to the same dead end.

Earlier in our conversation, Denys had pointed out to me that the historical data—from Greek and Roman civilizations through to developed countries in the late nineteenth and twentieth centuries—show remarkable consistency in suicide rates over time, despite an explosion of technological change. He'd then given me a macabre overview of how the means of suicide, by contrast, changed with the times. In England and Wales, hanging and drowning were common in the 1800s, but were progressively replaced by drugs and gassing. In Japan, hanging prevailed until 1950, after which pills and poison became the primary method.

"What this says to me," Denys explained. "Is that the motives are more constant than the means."

*

To say that suicide is part of our species' DNA doesn't mean that learning isn't involved. Like any complex behavior, pitting nature against nurture makes for a false dichotomy. When a previously unknown suicide method is introduced to a society, for example, the problem of social contagion can become an urgent priority.

In Hong Kong, the busy forests of skyscrapers and the deep-water ports claim a lot of jumpers. It's been that way for a long time. But ever since that part of the world found out about Jessica Choi yuk-Chun, officials have been scrambling to restrict the public's access to charcoal. In November 1998, alone at her home in an upscale suburb of Hong Kong, the young insurance executive meticulously sealed off all of the openings to her bedroom, tossed a lit match into a charcoal grill that she'd set up in the middle of the room, and crawled beneath her bedsheets to die quietly of carbon monoxide poisoning.

To this day, nobody knows how the shy businesswoman came up with this unusual manner of death. Few had ever heard of such a thing. Whatever her inspiration, though, once the Chinese media published glamorized accounts of her suicide, complete with illustrations and depicting the method as a painless, discreet new way to die,* "charcoal-burning suicide" became a major problem. Within a few short years, it was the second most common way to kill oneself in all of Hong Kong (after only jumping), and by the end of the next decade, it had become a full-blown epidemic.

Today charcoal-burning is a leading cause of death in several other Asian societies, too, including Macau, Taiwan, and Japan. It's not so easy to buy packets of charcoal anymore in these places. Even if you're just planning a weekend barbeque with the family, the clerk will probably look at you askance as he removes a container of charcoal from a locked cabinet and then gently taps the government's

*Since it's highly fatal, there aren't many people around who've survived an attempt at charcoal-burning suicide. Within that small group, however, there are mixed accounts of what it feels like. Some claim there was no discomfort at all, while others were surprised to find it wasn't quite as serene as they anticipated. "It is a suffocating experience which is extremely unpleasant," according to one expert. "The process involves displacing oxygen, almost like being strangled."

warning message that now comes printed on every box: "Cherish your life. We're here to listen."*

Cases such as that of Choi yuk-Chun remind us that social learning exerts tremendous pressure on all aspects of suicide, from rates to methods to cultural attitudes. Yet the similarities across vast reaches of place and time also tell us that we ignore our species' evolved psychology at our own peril.

"Any general theory of suicide," Denys wrote many years ago, "must account for the biological anomaly that this behavior presents."

And he's still right.

Yet I'm not convinced that suicide is an evolved adaptation.

Don't get me wrong. I'm not unconvinced, either. It's just that, like others who've carefully reviewed this literature, it seems to me that some things fit, and others don't. It's best, I think, to reserve judgment until more solid hypothesis-driven data are in hand.

Until that time, though, Denys's evolutionary approach challenges us to step back—*way, way* back—and see suicide from a very different angle than the conventional medical one that we've grown so accustomed to. And whether his theory proves to be right or wrong, I think there's tremendous value, and courage, just in doing that.

"If you could go back in time," I asked Denys, "climb up on that bridge in Halifax where Gregory is about to jump, and speak to him knowing what you know now, what would you say?"

"Come back to the family," he said, tearing up. "Find work. Start again. Just, please, come home."

*When it comes to the spread of popular new methods, sociologists have noticed a worrying trend of suicide contagion across cultural borders. The internet has eroded what were once informational boundaries due to language and geography, and global online buzz among the depressed is a real issue. Consider Boston lead singer Brad Delp, who almost certainly read about charcoal-burning suicide on the Web before lighting two charcoal grills in the bathtub of his New Hampshire home in 2007. When first responders arrived at the scene, they found Delp lifeless on a pillow. Affixed with a paper clip to the collar of his shirt was a simple note: "*J'ai une âme solitaire*" (I am a lonely soul).

4

hacking the suicidal mind

There are moments, perhaps not known to everyone, when a man may be nearly crushed by the terrible awareness of his isolation from every other human being.

T. S. Eliot, "Literature and the Modern World" (1935)

Have you ever done something that you immediately regretted doing?

I don't mean devouring a whole tub of Ben & Jerry's in one sitting or being a little too blunt with your feedback when your friend asks you the night before her wedding if you think she's overdone it with the self-tanning product. Rather, I mean something more along the lines of, just for example, mailing a melodramatic love letter to the object of your infatuation in which you confess your endless longing and burning gay desire for him, without knowing—or in fact having any reason whatsoever to believe or even suspect—that this person, this perfect young Adonis whose initials you've secretly been scribbling in your notebook every day for the past two years, is anything but straight.

Okay, fine, maybe it's just me then.

Nevertheless, this is the predicament in which I found myself as an easily mortified, closeted high school junior who made a boldly romantic move in a time of lucid stupidity. I knew instantly the gravity of what I'd done upon slipping the envelope into the irretrievable black void of the mailbox. Why did I do it then, you ask? I can only assume it was some strange alchemy: the naïveté of youth, perhaps, combined with the deluding power of lust and my growing distaste for the lies I'd by then long been spinning about my sexuality.*

Looking back, it sounds like an eye-rolling parody. But what might make for a formulaic secondary plot in some crappy teen romcom today was in fact a nightmarish situation for me in small-town Ohio in 1992. For four torturous, relentlessly worrisome days, I experienced a level of anxiety, hopeless despair, and shame that I'd never known before. By that evening, I'd convinced myself that the other boy was undoubtedly straight and, having foolishly entrusted him with an amount of discretion impossible for our age, I braced myself for my heterosexual mask being ripped off in the most painful of ways. Every waking moment, my head was filled with visions of what was to come: the onslaught of homophobic slurs in the hallways, my overnight metamorphosis into a pariah, the excruciatingly awkward conversations with my family. "I'm calling about your son," I imagined his irate father speaking to mine over the phone. "Do you know about this disgusting letter he wrote to my kid? No? Here, I'll read it to you . . ."

I was a pitiable creature. I found myself at the top of a hundred-foot reservoir near my house, where I stared down at the valley of the massive dam and visualized, over and over, my body falling fast, whistling in the wind, thudding into the concrete sides. What prompted these suicide-fantasy excursions was a craving for qui-

*According to the late psychologist Dorothy Tennov, I was suffering from a telltale case of *limerence*—a neologism that means intense emotional and sexual attraction for a desired romantic partner. Here are its key symptoms: intrusive thinking about the person; a yearning for the other person to reciprocate the feelings; the inability to have such feelings for any other person; a fear of rejection; heightened sensitivity to signs of interest on the other's part; and the tendency to dwell on the person's positive characteristics and avoid the negative. Tennov believed that nearly all adolescents are stricken with a hobbling bout of limerence at some point in their burgeoning sex lives. Also known as "passionate love" and "infatuation," it's strikingly common. Psychologist Craig Hill and his colleagues found that such experiences tend to occur primarily between the ages of sixteen to twenty years. Although there are no differences between the sexes in having an infatuation, males are significantly more likely than females to have a meaningful unrequited lustful relationship.

escence, a desperate muting of my furiously buzzing thoughts. I wanted out of my head.

Imagine my relief, then, when the following Monday, while skipping school for fear of reaping what I'd sown at the post office the Thursday before, I discovered my own unopened letter in the mailbox, with a gloriously bureaucratic "Return to Sender/Wrong Address" stamped in faded red ink. Somehow, my diligent cowardly subconscious had managed to subvert my carnal brain by writing my own zip code instead of his.*

Yet in that four-day interval preceding this unlikely twist of fate, I was in an altered state of consciousness, one that I now know to be paradigmatic of the suicidal mind. For example, everything—or, rather, everyone—seemed so incredibly distant. That's not just a figure of speech. It was as though I were seeing the world through the wrong end of a pair of binoculars. And I had a strange compulsion to busy myself in homework, which is something that I assure you was an unfamiliar drive for me at the time. No matter how dull the content, the act of reading replaced my own rancid thoughts with someone else's. I read voraciously. It didn't matter what—mostly junk novels, in fact. Words simply served as a sort of emotional prosthetic; the absorbing thoughts of strangers were being pulled taut over my exhausting ruminations like a seamless glove being stretched over a jarringly disfigured hand.

If I had easier access to drugs or alcohol, I'd have eagerly indulged these self-medicating agents, anything to plaster those biting fears of

*That wasn't the end of the story. In my twenties, he continued to haunt my dreams, and the only way I knew to exorcise him from my obsessive thoughts was to write to him again, finally confessing all. I quoted Pascal: "The heart has its reasons which reason knows nothing of." I never heard back. Then, a decade after *that,* I emailed him on the eve of our twentieth reunion apologizing profusely for my rudely barging in on what was—I now realized—a very heterosexual suburban life. This time, he wrote back: "Dear Jesse, I am really glad to hear you are doing well and I have always felt bad for not responding to your letter. I certainly appreciated the courage it took you to reach out and express yourself so openly. I was not offended or upset by it in any way and I actually used it several times to try to make [my wife] jealous. No luck. Please know that I harbor no bad feelings and wish you only the best." I still dream of him, alas.

my imminent exposure. Sleep was intoxicating. I've encountered it many times since, but this was the first time in my life when I experienced what I've come to call the "bleary bliss phenomenon": that fleeting four or five seconds immediately upon waking when your mind is blissfully free of whatever it is that worried you so as you drifted off to sleep, a brief slice of time in which the heavenliness of the banal is revealed only by the abrupt return of a piercing disquiet.

What is the nature of that disquiet, and how does it differ from our day-to-day worries? The lonely English vicar Robert Burton described the suicidal frame of mind in his early dystopian psychological treatise, *The Anatomy of Melancholy,* in these nightmarish terms:

> In such sort doth the torture and extremity of his misery torment him, that he can take no pleasure in his life, but it is in a manner enforced to offer violence unto himself, to be freed from his present insufferable pains. So some (saith Fracastorius) "in fury, but most in despair, sorrow, fear, and out of anguish and vexation of their souls, offer violence to themselves: for their life is unhappy and miserable. They can take no rest in the night, nor sleep, or if they do slumber, fearful dreams astonish them." In the daytime they are affrighted still by some terrible object, and torn in pieces with suspicion, fear, sorrow, discontents, cares, shame, anguish, &c. as so many wild horses, that they cannot be quiet an hour, a minute of time, but even against their wills they are intent, and still thinking of it, they cannot forget it, it grinds their souls day and night, they are perpetually tormented, a burden to themselves . . . they cannot eat, drink or sleep.

Yeah—that about sums it up.

*

Many years later, when in grad school I happened upon an old article titled "Suicide as Escape from Self," written by the social psychologist Roy Baumeister, I realized just how close I'd actually come to the edge of that damn dam.

Now based at the University of Queensland in Brisbane, Australia, the Cleveland native is something of a legend to those of my own academic generation. Roy's provocative, often controversial, work on everything from willpower to sex to self-esteem and violence is among the most cited research in these already crowded areas. But to me, it's his incisive analysis of what it *feels like* to want to kill yourself that is his most significant contribution.

If you picture a weathered German sea captain, you'll have a pretty good image of Roy. His inscrutable face is ringed by a shock of blond-white hair, and a matching beard is for his occasional stroking as he considers your inane questions, such as, for instance, my always awkward icebreaker: "So how'd you get into suicide?"

"A series of accidents," Roy told me. "I started to write a book on the meaning of life. I was reading the suicide literature thinking it would help with that, and actually it didn't help all that much. But I kept finding really interesting things, so it became an intriguing problem to me. Yeah, some of the stuff wasn't that great. You can't do controlled experiments to see which people kill themselves. But there's tons of information, whole journals devoted to the subject."

"Interesting," I said.

Secretly, though, I suppose I wanted something more from Roy's answer than suicide simply being an academic drill or a riddle for him to solve at the height of his sterling career, because, to me, his breakdown of the suicidal mind offered a kind of preternatural insight into my own experiences, reminiscent of what I'd felt upon first reading, say, Dostoevsky or Faulkner. At the time, I couldn't help but think to myself, "Ah, yes, this writer *knows*."

(It's a prejudice of mine, I admit it, but it's difficult for me to fully trust anyone for whom the thought of taking one's own life hasn't alighted at least briefly, enticingly, on their thoughts; it suggests to me they probably haven't suffered enough to be fully human. Roy's remarkably clear plotting out of this singularly hellish state, an altered state of consciousness with its own cognitive biases and pigheaded whirls of emotion, made me feel, then and now, less alone. That's a feeling, of solidarity with the solitary, which I'd like to share with you as well.)

I asked Roy if he'd ever personally occupied the very dark places he describes in his model of the suicidal mind. "As someone who's been there," I shared with him, "it felt so uncannily accurate to me."

"No, not at all," he told me. "I can't say I've ever been suicidal."

His unexpected answer bewildered me in a way similar to how I'd felt when I discovered that the gentle, soft-spoken veterinarian I'd been taking my pets to for years mentioned offhandedly that he was an avid hunter. It's presumptuous of me to try to fit anyone in an emotional box simply because of that person's chosen profession or knowledge.

But still.

"I really enjoyed suicide," Roy said. "It was just so fascinating to work with, and then I got a paper in *Psychological Review,* which was a big coup." (That's the same piece, incidentally, which had been so meaningful to me. It's the only substantive writing on suicide Roy ever published.) "So I think I have more positive associations to suicide probably than anybody else in that research area," he continued. "For a lot of people, it has an intense personal meaning, but it didn't have any personal significance to me. It was just this intellectual puzzle."

But surely, I thought, there must be something in his past that made suicide more than just an "intellectual puzzle."

Gingerly, I poked. "So, thinking back, to your school days, or to any friends or work colleagues that might have taken their own lives, or even just had a history with suicidal thoughts . . . ?"

He looked genuinely stumped.

"About three years ago a nephew committed suicide," Roy seemed to suddenly recall. "But that may be the only one."

Still, maybe his nephew's death had deeply affected Roy somehow, or had him revisiting his earlier ideas about suicide.

"Were you close to him? Or was it more—"

"No—" he cut me off. "I wasn't."

So much for that.

Before moving on with our conversation about suicide, then, I was forced to abandon this tack. Perhaps Roy's genius simply lies in his ability to piece together the most elusive mental operations, even ones that exist for him only in impersonal theoretical terms. Maybe, in fact, it's their very foreignness to him that allows him to see them so clearly.

✷

The best way to approach Roy's model is to look at it as a series of steps, or stages, in which the individual moves successively from one degree of suicidality on to the next, with each step becoming increasingly dangerous. There are six steps in all.

"Can you disrupt the process?" I asked Roy, just to be sure we're on the same page with the more practical value of his stage-theory account. "Can you pull yourself out of it once you've begun cycling through these stages?"

"Yes," he confirmed. "The whole point is that you can get off the path to suicide at each step. If you can manage to get rid of the emotional distress in other ways, then you don't go on to suicide."

Keep in mind that the main purpose of this exercise is to help you to recognize distinct landmarks in your own suicidal mind or that of someone you love. It's far from obvious that all suicidal people even realize they're suicidal, or at least know when they've begun to veer off in this direction. Had someone asked me at my lowest if I was suicidal, I'd have said no. In doing so, I'd have been lying to myself even more than to the person asking. Like being gay—a part of my identity that would take years for me to acknowledge, let alone accept—being "suicidal," as a word, as a concept, was something out of my orbit, an aspect of other people's lives. But mine? I'm not one of *those* people, said the pot to the kettle.

If we're all potentially suicidal, as I suggested in the previous chapter, then it's not just the outwardly "mentally ill" who are at risk, and it would be a grave mistake to regard ourselves as being entirely safe from ever meeting death at our own hands.

One other thing before we get started: Although they're meant to be seen as progressive, there's an elliptical aspect to Roy's stages as well. They often overlap. Just because you're in Stage 4, for instance, doesn't mean that the defining features of Stage 2 are no longer occurring, to some degree. (And in rare cases, they can even happen all at once.*) One can go back and forth along the trail of suicidal thinking, in other words, and in principle turn around at any time. But the further you go, the harder that becomes.

Stage 1: Falling Short of Expectations

One of the more surprising things about suicide is that most people who kill themselves have actually lived better-than-average lives. When we get a little too accustomed to uneventful and pleasant conditions, warns Roy, a sudden, precipitous drop in our standard of living can dangerously disorient us. It's the law of social gravity: Compared to the guy who's been living his whole life down there, hitting rock bottom is going to hurt a lot more for the person taking a tumble off the side of a mountain. We're all downwardly mobile to some extent, but it's the *magnitude* of the discrepancy between our personal standards and our current life situation that plays a role in the presuicidal process. An experience that to many of us wouldn't seem so bad, or at least certainly not something to end one's life over, to others makes for an unlivable existence. This is because that individual has unrealistic—or unsustainable—criteria for their success.

Consider the recent death of former professional rugby player Dan Vickerman, who played for the Wallabies and several other Australian teams from 2002 to 2011. After a persistent leg injury began slowing him down on the field, he was forced to begin thinking about life after what had been a comet-like rise in the sport. Anticipating this, Vickerman, an intense young South African with a perfectionist streak, left Australia in 2007 and spent three years studying at Cambridge University, where he played for the Northampton Saints while earning a degree in land economics.

The future, however, came faster than Vickerman would have liked. Not long after mounting a brief comeback with the Wallabies

*In September 2017, for instance, a thirty-eight-year-old man in South Carolina found his two-year-old son dead; the boy had stumbled on his dad's loaded weapon and accidentally shot himself in the head. Overwhelmed, the man then turned the gun on himself, dying in the same manner.

at the 2011 Rugby World Cup quarterfinal, then an unremarkable season with another team, he officially retired from the game, beginning his transition "from elite athlete to regular human," as the media reported.

From the outside looking in, Vickerman appeared to be doing this just fine. He landed a coveted job with a big hedge fund company in Sydney, dabbled in real estate, and became chair of the Australian Rugby Union's committee dedicated to player development. In his spare time, he played on an "old boys' team" called the Silver Foxes with other former pros. Secretly, though, the husband and father of two small children confided to friends how he was struggling with this new pedestrian life. On the night of February 18, 2017, police were called to the New South Wales home of the thirty-seven-year-old, where, sadly, he'd killed himself.

One of his friends, Gregg Mumm, a consultant to professional athletes, thinks Vickerman's death should serve as a wake-up call for the industry to begin acknowledging the challenges faced by people too old to continue their lives as sports stars, but too young to live placidly in the regular world. "What happens," asked Mumm, "when the stadium lights are turned off on [their] careers and the fans find someone new to cheer for?"

Of course, it's not just professional athletes who feel this loss of purpose and meaning in the aftermath of a successful career. Among white males in the U.S., suicides begin to noticeably creep up around the usual retirement age of sixty-five. More generally, Vickerman's death exemplifies a key aspect of Roy's stage-theory model of suicide, which is that those most at risk have experienced some recent setback or find themselves caught in that dark membranous space dividing a happy past from (what they perceive to be) a hopeless future.

In his *Psychological Review* article, Roy amassed a mountain of epidemiological data to support the claim that suicidal thinking is precipitated by events that fall short of high standards and expectations.* Sometimes, as in Vickerman's case, these expectations are

*Suicide rates are higher in developed nations than in less prosperous ones; higher in U.S. states with a better quality of life; higher in societies that endorse individual

owed to our own past achievements, but they can also be the result of what Roy refers to as "chronically favorable circumstances." Simply being poor isn't a risk factor for suicide. But going suddenly from relative prosperity to poverty is. Likewise, being a lifelong single person isn't a risk factor, but the abrupt shift from being married to being single places one at significant risk of suicide. Most suicides in jails and mental hospitals occur within the first month of confinement, during the person's initial adjustment to this sterile new existence.

Dealing with unreasonable or impossible external demands can be our downfall as well. When others place their trust in us in ways that are overwhelming, the fear of letting them down can be crushing. This may be one reason why students who take their own lives often have a glowing record of academic achievement and parents with high expectations, but in the semester preceding their suicides, their grades fall.

Not long ago, for instance, a young man from India—let's call him "Malik"—reached out to me through Facebook:

> I need help. I read an article you wrote on suicide. It mentioned that you were once a suicidal adolescent and I hope that you would understand my situation. I'm severely depressed and suicidal and literally crying right now. I'm nearly 20 years old and on the verge of killing myself. Help me.

There's a seven-hour time difference between New Zealand and India, so when I saw that Malik's message had been written the previous day, I worried that I might already be too late. I was hugely relieved to see that he'd sent another note just a few hours after the first:

---

freedoms; higher in areas with better weather (in areas with seasonal change, they are higher during the warmer seasons). They're also lowest on Fridays and highest on Mondays; and they drop just before the major holidays before spiking sharply immediately after the holidays. Baumeister interprets these patterns as consistent with the idea that people's high expectations for holidays and weekends materialize, after the fact, as bitter disappointments.

> I'm sorry to disturb you but I get sudden mood changes and got suicidal. I had to vent it out. . . . I'm feeling better at the moment (talked to my mom about it). I'm at a low point in life. I'm doing my degree in computer engineering but failed all subjects my first year. I had always been a high achiever and continuous failures kept on dragging me down spiritually. I stressed myself too much about it because of all the pressure on me. I'm the eldest brother and you must be aware of my country economically. I've had too many responsibilities thrust upon me. I just wanted to give up at the time.

Malik's situation is an all-too-common one on university campuses. Here's how a middle-aged woman from New York reflected back upon the suicide of her older brother thirty years earlier:

> He was good-looking, he was popular, he was off at college and he just couldn't get his act together. He was so smart . . . [but] his addictions just kept getting worse, so he was just kind of failing out. He dropped out of college and then he [was] working as a manual laborer and I think he just felt depressed, he felt like a failure and he just didn't see his way out of it. You know, there is kind of the family dysfunction that created him like that, and then also the family expectations that were created even in the midst of this dysfunction that we were going to succeed and do great things, you know?*

Stage 2: Attributions to Self

If all it took to send us on a suicidal tailspin were the vicissitudes of fortune, everyone would end up killing themselves. As Roy likes to say, "Bad things happen." He's right; that's just part of the bargain for getting to be alive. According to the second stage of his model, rather, it's when you blame yourself for the unfortunate events from

*There's a deep-seated hypocrisy behind this central message that we impose on children, this fetishizing of greatness. We convey to them that it's bad to be "just" an average person. Or at best, we're telling ourselves that we should never be content with being average people, which is, for better or worse, the only thing that most of us can ever possibly hope to be.

the first stage that you begin to amble further and further along the suicidal trail.

When we loathe ourselves over the particular trouble we're in, that's a red alert. Across cultures, self-blame or "condemnation of the self" is a common denominator in suicides. It's not simply low self-esteem that puts the individual at risk, but a *recent* demonization of the self in response to whatever has gone so wrong (or threatens to go wrong).

There's some irony here. People who've always thought poorly of themselves, possessing a sort of innately gloomy personality that, even since they were little, has made them less than thrilled to be who they are, actually enjoy a protective buffer against suicide. This is because, while they're indeed poisonously self-critical, they tend to hold the rest of humanity in similarly low regard. We all know the type. They're called misanthropes.

The suicidal person, by contrast, hates him- or herself too, but they typically suffer the erroneous impression that other people are mostly good, while they themselves are bad. As we saw in chapter 2, to feel suicidal is to feel the unbearable weight of other people's thoughts bearing down on us, even if only the faceless and unforgiving society at large that we've come to internalize as our own personal judge, not altogether dissimilar to Freud's famous superego. In sociology, this is also in line with the concept of the "looking-glass self": our self-image is a product of how we believe other people see us. Because these beliefs are based on assumptions that we're making about others' evaluations of us, and these assumptions are often wrong, our self-image is intrinsically askew. For healthy people, it tends to be askew in a flattering kind of way.* Most of us see ourselves

*In one study, psychologists Nicholas Epley and Erin Whitchurch took photos of undergrads with a neutral facial expression, invited these same students back to the lab two weeks later, and asked them to identify their actual face out of an assortment of eleven possible images. But here's the clever part. These other images were in fact the actual face morphed to varying degrees with either an extremely attractive composite face or unattractive faces with craniofacial syndrome. The results? On a variety of different measures, the participants were significantly more likely to choose a more attractive morphed face as being their actual face than even their non-morphed actual face! The authors conclude, "It is perhaps of little wonder, then, that people so rarely

as being more physically attractive, smarter, more likable, and funnier than other people really see us. As we discovered in the previous chapter, however, being depressed often makes this reflected self-image more accurate, and unflatteringly so.

And because depression also makes us more sensitive to signs of social rejection, we begin overestimating how much people actually care about our shortcomings. Of course, we really are being judged by others much of the time; it's just that other people don't actually care about our foibles and flaws as much as we think they do. They're too busy worrying about what others think of their own. Nevertheless, our inevitable failures, and sometimes our catastrophic missteps, are made all the more treacherous when coupled with this tendency to amplify in our own depressed minds other people's awareness of our undesirable qualities.

That always fragile looking-glass self—an image reflecting feelings of worthlessness, shame, guilt, inadequacy, or being exposed, humiliated, and rejected—leads suicidal people to dislike themselves in a manner that makes them feel, quite literally, unworthy of inhabiting this world. It's as though they are seeing themselves for the first time and they've caught an especially alarming sight of a face prone either to great failure or great evil. In their mind, this shocking visage shatters a long-standing illusion of their reflected public identity. Now, the "real" self is seen for what it is: enduringly undesirable, irreparably broken, and embarrassingly flawed. There's no hope for change: "I" am simply rotten to the core.

I know all too well this feeling of being cleaved off from an idealized humanity. Perhaps the clearest example of this coincides with that wayward love-letter episode I shared at the start of this chapter. As I was holed up in my bedroom listening to the radio in a futile attempt to zone out, it suddenly struck me that, although I liked them well enough, none of these pop songs were meant for me. It was as though I were a filthy imp eavesdropping on angels. Surely, if only these singers knew what I'd done, I thought to myself, they'd

---

seem to like the photographs taken of themselves. The image captured by the camera lens just doesn't match up to the image captured in the mind's eye."

say the same thing: "This is for *normal* people, not queers like you infatuated with other boys."

The sense of being isolated from good, regular folk in some unbridgeable way will attend many suicides. From our vantage point as decent, neutral observers, a rapprochement seemed more than possible. After someone takes their own life and the truth comes out, we're often left wondering why it was so hard for that person simply to share with trusted others what the problem was, exactly. But opening up was obviously so terrifying to the victim that suicide was seen as a less painful alterative than having that hard conversation. For me, at that time, I had a strong urge to make an unbidden confession about the letter, but I wasn't ready to come out of the closet—certainly not to my friends or parents, and I wasn't about to share the unspeakable details of my crisis with my, er, girlfriend. So instead, I was left in this state of mute limbo, and nobody can survive there for very long.

While it's usually far from being the case, many suicidal people feel as though their situation is uniquely insufferable, that nobody else in all of history has ever had to endure such singular monsters. Here's part of an email I received from one such reader:

> Every inch I climb turns into a three-foot slide. It's really bad. I even look at others who have killed themselves and say, "You ain't got nothing on me; I only wish I had your problems." I'm scared to talk to friends or family thinking they would turn me into some hospital or, worse, ignore me.

This person perceives his situation as being in a league of its own; at some level, he's aware of the absurdity in this reasoning, and yet he still can't help but feel jealous even of people who were in so much pain that they'd actually ended their lives, thinking they must have had it easy by comparison. At the same time, he's paralyzed into silence for fear of his loved ones' reactions and unwilling to communicate his unutterable worries.

In pondering this second stage, I've always had the niggling sense that "stepping off" the trail to suicide is probably much harder for some people than it is for others. I mentioned this to Roy. "On the one hand," I said, "obviously, it's best to learn from our mistakes and then move on to something else. But on the other hand, is not blaming themselves really a decision that people can make, or is it more of an immutable trait, with some people being inherently more likely to blame themselves than others?"

"Well, we can all learn to go easier on ourselves," Roy said, "but the latter is probably also true. One presently obvious example is President Trump, who seems so quick to blame anything that goes wrong on other people. It's hard to imagine him killing himself.* If you blame society—or anyone but yourself, really, such as the media—then you're not going to die by suicide because you're not experiencing the kind of emotional distress that leads to it."

We tend to think of psychiatrically disturbed people as delusional. But sometimes, they're the ones who are seeing things most clearly, and that's part of what is driving them mad. It could in fact be true that the self is to blame; but it can also be unhealthy, in the clinical sense of that term, to have such absolute clarity about one's exact role in the problems one is facing. Too much of a good thing is a disaster when it comes to mental delusions, in other words, but a pinch of it never hurt anyone. Quite the opposite, it seems.

Stage 3: High Self-Awareness

At the heart of Roy's theory is the idea that suicide is motivated by the desire to escape from an unpleasantly sharp self-awareness. When we're stuck in the self-destructive mind-set, we're egocentric, and other people seem impossibly far away. This isn't the sort of vain egocentrism that we'd normally associate with, say, a narcissist. It's an unwanted preoccupation with the self's flaws. That is, as a result of our ongoing and unflattering comparison of the self with personal standards, consciousness begins to encroach on us in an

*Incidentally, Roy is notoriously apolitical; in the past, he's written op-eds stating that, as a social psychologist, he is ethically obligated only to observe, not to vote.

all-absorbing way, which is very painful.* When your every waking thought is about how despicable, unlovable, or useless a person you are, consciousness is agonizing.

"I know that I get really, really self-conscious," one woman said when asked how she knows she's getting suicidal.

> I am normally not the least bit self-conscious. But [when I'm in this state,] I will isolate to where I don't want to leave my room. . . . I just cut people off and ignore the cell phone, ignore Facebook, and just don't want to talk to anybody. And I just want to sleep all the time because I don't even want to be in my own head. It is exhausting. I can remember being in my car and driving was exhausting, and even the energy of driving myself off the road was just too much. . . . And then I talked to other people who were suicidal, and they were like, "I am exhausted too. I know. It is draining."

"There's that old refrain," I mentioned to Roy, "that 'suicide is a selfish act.' Is it, in some ways, true?"

"I think it's said deliberately to pressure people to not commit suicide," Roy said. "But there's probably a sense in which it's correct . . . you're doing it to escape your own personal misery, and not all that concerned with the effects it will have on others, or that it will create difficulties for them."

This lack of concern, however, is not an intentional omission of others' feelings, but a feature of the distortions characterizing the suicidal mind. The person is experiencing a *cognitive inability*, or at least a temporarily weakened capacity, to empathize. They have trouble putting themselves into other people's shoes when thinking about how their death will impact the lives of everyone they care about, and who care about them. In an interview with Bonnie Scarth, a once-suicidal young woman expressed this clearly:

*Hence the meaning of the title for Kay Redfield Jamison's powerful scientific treatise on suicide, *Night Falls Fast: Understanding Suicide*.

> Most people's understanding of suicide is so backwards. My boyfriend made a comment the other day—we were watching TV—and he said, "If I was going to kill myself I would make sure there was a real big 'Fuck You' to someone who screwed me over," and I was like, "But you don't." He goes, "What do you mean?" I said "You don't. That is not what you care about when you feel that way." There is literally no concern for that . . . when you really want to kill yourself, you are cut off at that point from the rest of the world and you could not care less what is going on with somebody else. Your thought process is not "Wow, this is really going to screw [that person's] day up!"

How do researchers capture the state of high self-awareness in *currently* suicidal minds? This piquancy of egocentric thought is measurable, at least indirectly, by analyzing the language used in suicide notes. As Edwin Shneidman wrote in *The Suicidal Mind,* "Our best route to understanding suicide is not through the study of the structure of the brain, nor the study of social statistics, nor the study of mental diseases, but directly through the study of human emotions described in plain English, in the words of the suicidal person."

It may sound morbid, but the study of suicide notes is a longstanding tradition in psychological research. Over the past few decades alone, hundreds of such research articles have been published. These studies have addressed a wide range of hypotheses, but because they've yielded inconsistent findings, they also paint a confusing picture of the suicidal mind. This is especially the case when trying to reveal people's motivations for the act. Some who commit suicide may not even be aware of their own reasons, or haven't been completely honest in their farewell letters to the world.*

The most compelling studies, in my view, are those that use

*A good example comes from the sociologist Susanne Langer and her colleagues. The researchers describe how the suicide note written by one young man was rather nondescript, mentioning feelings of loneliness and emptiness as causing his suicide, while, in fact, "his file contained a memo inquiring about the state of an investigation regarding sexual offences the deceased had been accused of in an adjacent jurisdiction."

text-analysis programs enabling investigators to make exact counts of particular kinds of words. Compared to fake notes (written by non-suicidal participants asked to write their own hypothetical suicide notes), real suicide notes are notorious for containing an abundance of first-person singular pronouns such as "I" and "me." Psycholinguists believe this is indicative of high self-awareness. And unlike letters composed by people facing involuntary death, such as prisoners about to be executed, suicide note writers rarely use inclusive language such as "us" and "we." (This same pattern has also been found online by researchers analyzing posts on Twitter, Facebook, and other popular social media platforms made by people who shortly went on to kill themselves.) When they do mention significant others, they usually speak of them as being cut off, distant, separate, not understanding, or opposed. Friends and family, even a loving mother at arm's length, feel miles away.

The fact that only about 30 percent of those who take their own lives bother to leave a note, an astonishingly low figure given the impact on those left behind, is revealing in itself of this sense of feeling fundamentally split off from others.* Here's how one man, who'd attempted suicide and had been dealing with these feelings for most of his life, explained this socially disconnected state to Bonnie in her series of interviews on the subject:

> There were times when I felt like there was this wall between me and my family. I would be sitting at dinner and feel like there was just this clear glass wall between me and everybody else. I could just tell I felt so much worse than they knew.

*Numerous studies have examined the possibility of differences between those who leave notes and those who don't, and very few distinctions between these groups emerge. Perhaps the best study in this area was a large-scale study done in 2009 that included 621 suicides in Ohio. The researchers compared note writers to non-note writers on forty separate variables (such as race, day of the week of the suicide, age, prior suicidal history, giving clues to the impending suicide, depression, psychosis, and so on). They found only two differences: note writers were more likely to have lived alone and to have made prior threats of suicide.

Suicide, of course, isn't the only way to dull a feverishly high self-awareness.

"Are there any elements here that are familiar to you?" I asked Roy. "Maybe not in the direct sense of suicidal thought, but any other sort of escape tactics that you can say you're more intimately familiar with, as a universal human experience?"

"Alcohol is an escape from self. I've certainly had some familiarity with that," he said. This reminded me of something that the poet Charles Bukowski once observed. "I have the feeling," Bukowski said, "that drinking is a form of suicide where you're allowed to return to life and begin all over the next day. It's like killing yourself, and then you're reborn. I guess I've lived about ten or fifteen thousand lives now."

A constant chasing of the dragon through drugs and alcohol is a daunting task that few can endure forever, however, and bouts of excess will be punctuated not only by sober moments refocusing our minds on whatever drove us to escape in the first place, but now the guilt and shame of chemical addictions and all the messy details of a life in gross disorder. "The important thing was not to stop," wrote Al Alvarez in *The Savage God* of the period leading up to his suicide attempt. "In this way, I got through a bottle of whiskey a day, and a good deal of wine and beer. . . . Anything not to stop, think, feel. The tension was so great that, without the booze, I would have splintered into sharp fragments."

The causal sequence can be difficult to discern: is it, for instance, suicidal feelings that drive one to become an alcoholic, or has the person become suicidal only because they're in the merciless grip of their dependency? Perhaps in the end, the directionality matters less than the common denominator, which is the motive to escape from the self. "I suspect drinking invokes an animal level of awareness," Roy told me. "When you're intoxicated, there's no differentiation of self from others, a limited recognition of separate minds. Alcohol moves you back closer to that animal-like state of consciousness."

Once, standing on a busy city street corner in the shadow of Auckland's Sky Tower and peering up at the bungee jumpers, my attention was drawn instead to street level, where a pair of jovial homeless men leaned against the wall of a liquor store. Despite their squalid state, there they were, lost in joyous banter, ribbing each other through inebriated peals of laughter while breezily watching the world and all its funny folk go by. For that splendid moment in time, perhaps they were, as Bukowski said, dead; made invisible by drink, they were untroubled phantoms enjoying us sharp-minded beasts in our cage.

Stage 4: Negative Affect

When David Foster Wallace, author of the modern classic *Infinite Jest,* decided to take his own life at the age of forty-six on September 12, 2008, he knew his body would rebel. So before standing on a garden chair in his cluttered California garage and slipping his neck into a makeshift noose—a leather belt—he bound his wrists together with duct tape to keep from instinctively trying to get free. Only then did he kick the chair out from beneath himself.

Terms such as "brilliant" and "creative genius" are limp compliments for a writer with Wallace's inimitable talents. His fluid, postmodernist style achieved soaring literary heights. Yet what few but those closest to him knew was that the surrealist terrors infesting his characters' lives were in fact all too familiar to him. He'd been grappling with anxiety attacks and serious depression since his school days, hospitalized for suicidal drug overdoses, had sampled almost every antipsychotic and antidepressant on the market, and twice allowed himself to be subjected to electroconvulsive shock treatments on the advice of his psychiatrists. And still his demons, most of which revolved around his tortured writing process, just would not abate.

Here's how, in *Infinite Jest,* Wallace described the compulsion for suicide:

> The so-called "psychotically depressed" person who tries to kill herself doesn't do so out of quote "hopelessness" or any abstract conviction that life's assets and debits do not square. And surely not because death seems suddenly appealing. The person in whom Its invisible agony reaches a certain unendurable level will kill herself

> the same way a trapped person will eventually jump from the window of a burning high-rise. Make no mistake about people who leap from burning windows. Their terror of falling from a great height is still just as great as it would be for you or me standing speculatively at the same window just checking out the view; i.e. the fear of falling remains a constant. The variable here is the other terror, the fire's flames: when the flames get close enough, falling to death becomes the slightly less terrible of two terrors. It's not desiring the fall; it's terror of the flames. And yet nobody down on the sidewalk, looking up and yelling "Don't!" and "Hang on!", can understand the jump. Not really. You'd have to have personally been trapped and felt flames to really understand a terror way beyond falling.

Anxiety—which can be experienced as guilt, self-blame, fear of ostracism, and perhaps, above all else, *worry*—seems to be a factor in the majority of suicides. Like those menacing flames starting to lick at the suicidal jumper in Wallace's analogy, something bad, something more terrifying than death, even, is felt to be closing in. And the person can see no way out. The fact that incredible physical pain is more tolerable by comparison helps us to see just how debilitating this degree of psychological suffering can be.

The British psychoanalyst Phil Mollon offers another useful metaphor, which is that the person in a suicidal state, particularly when it's induced by unbearable shame, is like a fox that has been forcibly dug out of its lair and surrounded by a whole pack of hounds that has exhausted its capacity to flee and hide. Roy's escape theory also helps us to understand self-harming behaviors, such as "cutting" in young people, which is essentially a way of forcing the mind to focus on input coming from physical pain receptors, thereby drawing it away, even if only temporarily, from its ferocious fretting over perceived social problems. One of Shneidman's patients, an anorexic college student, describes this self-harming response to her boyfriend dumping her. "I could not handle the overflowing waves of pain that washed through my body immediately after the break up. I had never felt such intense pain and I could not handle it. . . ."

> I was alone at home and ran desperately around, panicked over the flood of emotions that was traveling through my body. I ended up taking the kitchen knife into my room and cutting myself, slashes all along my arms. The physical pain let me pull my attention off the emotional agony, and I just concentrated on not letting the blood spill over onto the carpet. That day I clearly remember wanting to die.

And a teenage girl, who went on to kill herself, describes her reasons for cutting even more clearly in a journal entry just a month before her death. She recounts how she tried to explain the perplexing act of self-harm to her counselor. "I reinforced that it is one of the hardest things for anyone to understand unless they have done it themselves," she wrote.

> I tried to use the comparison of it being a coping mechanism like how anyone would turn to drinking or cigarettes when they experience emotional distress. The physical pain is worth the escape. The pain itself is the escape. It's like a device that turns mental pain into physical pain. The pain distracts you. It can either give you a break, or make you feel something. It depends on whether you feel numb from dissociation and you want to remind yourself that you're still alive, or that you feel too alive and want the pain to make you numb. . . . But the result is always the same. The pain distracts you and is a form of release. The endorphins momentarily fix your problems. Seeing the crimson makes it seem like the cut wasn't a worthless scratch. That it wasn't for nothing. But in the end, you're always left with the scars you have to hide.*

Psychodynamic theorists have postulated that suicidal people feel guilty about something and therefore seek punishment, a sort of rage turned inward, and thus suicide is a type of self-imposed death penalty. But Roy largely rejects this interpretation. Rather,

*From the journal of Victoria ("Vic") McLeod. We'll return to Vic's story in the next chapter.

in his escape model, the appeal of suicide is loss of consciousness, and thus the end of psychological pain—"negative affect"—being experienced. Unable to find peace of mind, we strive for the peace of mindlessness.

"How do you see your escape theory applying to those so-called cry-for-help or attention-seeking types of suicide attempts," I asked Roy, "where the person doesn't really want to die? Are they different from people who use more guaranteed lethal means?"

"They wouldn't be entirely exclusive," Roy says, "because if you do a fake suicide attempt and then you sort of get taken out of your life for a while and put in the hospital and people take care of you, it does bring about a kind of escape, as well as you getting help."

The account of one Japanese woman who'd survived a suicide attempt would appear to support Roy's line of reasoning. "When I think about it," she said, "it is not that I really wanted to die. I just wanted to pause from living. To have died or not to have died, either would have been all right." Like using drugs to flee consciousness, however, suicide attempts are an anesthesia that eventually wears off. And again, a history of self-harming behaviors or prior attempts—for many people, both—places one at much higher risk of an actual suicide, sooner or later.

The most important point to stress with regards to this fourth step is that the majority of suicides are driven by a need to escape from immense and ongoing negative affect (that aptly named "psychache," if you recall). They do not follow a cool-headed stream of Socratic dialogue. A suicidal person doesn't idly contemplate "to be, or not to be," nor do they ask, as someone once falsely accused Camus of writing, "Should I kill myself, or have a cup of coffee?" When someone you care about is in this state, trying to use reason and logic with them can be as useful as advising a person with a compound fracture of the leg to just walk it off, or telling a schizophrenic patient that it's all just in their head.

For the truly suicidal, consciousness is incapacitating.

Stage 5: Cognitive Deconstruction

The fifth step in the escape theory is perhaps the most intriguing, from a psychological perspective, because it illustrates just how scarily distinct the suicidal mind is from that of our everyday cognition. With cognitive deconstruction, a concept originally proposed by social psychologists Robin Vallacher and Daniel Wegner, the outside world becomes a much simpler affair in our heads—and not in a good way.

Cognitive deconstruction is just what it sounds like. Things are cognitively broken down into increasingly low-level, basic elements. As part of this process, the *time perspective* of suicidal people is affected. Time crawls. As Roy writes: "Suicidal people have an aversive or anxious awareness of the recent past (and possibly the future too), from which they seek to escape into a narrow, unemotional focus on the present moment."

In one study, when compared to control groups, suicidal participants significantly overestimated the passage of experimentally controlled intervals of time by a large amount. "Thus, suicidal people resemble acutely bored people," Roy explained. "The present seems endless and vaguely unpleasant, and whenever one checks the clock, one is surprised at how little time has actually elapsed."

This warped sense of time is difficult to articulate. "I was counting the hours and the days, and then it was the minutes," one woman who'd attempted suicide said to Bonnie, "and every minute was torture. It was so painful! I would go home and it would just be like . . . it was like I am going to bed now and watch TV and your life just *slowly drips out*."

Roy believes that this kind of temporal winnowing, in which the mind is constricted to a molasses-like present where life "slowly drips out," is actually a defensive mechanism helping the person to stop dwelling on past failures and from worrying about an intolerable, hopeless future. As a consequence of deconstructed thought—of this occupation of the meaningless moment—those negative emotions from the previous stage are to some degree assuaged. This is why so many suicides are preceded not by outbursts or fits of rage, as one might assume, but instead more of a shell-shocked state of flat affect . . . of ennui.

The sterile present is no place to live and brings with it its own suffering. "[It] comes to resemble the diabolic discomfort of being imprisoned in a fiercely overheated room," wrote William Styron in his memoir *Darkness Visible*. "And because no breeze stirs this cauldron, because there is no escape from this smothering confinement, it is entirely natural that the victim begins to think ceaselessly of oblivion."

One man who eventually killed himself spoke of feeling "emotionally constipated."

Another aspect of the suicidal person's cognitive deconstruction is a dramatic increase in concrete thought. Like the intrusively high self-awareness discussed earlier, this *concreteness* is often conveyed in suicide notes. Several review articles have reported a paucity of "thinking words" in suicide notes. Instead, they more often include mundane details, such as "Don't forget to feed the cat" or "Remember to take care of the electric bill." One old study even found that genuine suicide notes contained more references to concrete objects in the environment—broken refrigerators, empty beer bottles, unpaid bills left on the kitchen table—than did pretend ones.

Suicide notes tend to be punchy and to the point, containing an average of only 150 words. "The typical purpose of the suicide note," writes one team of investigators, "is to convey final instructions and factual information . . . rather than to provide an existential account of the futility of one's life." Real suicide notes are usually void of contemplative thoughts, whereas fake suicide notes include more abstract or philosophical wordage ("someday you'll understand how much I loved you" or "teach my son to be a good man").

What this cognitive shift to concrete thinking reflects is the brain's attempt to slip into idle mental labor, thereby avoiding the overwhelming emotions that we've been describing. "I make lists," one woman told Bonnie, "because it pulls me away from that feeling of hopelessness and that 'oh my God I want to die' feeling. . . . I go from 30 to 120 in mere seconds, and when I am at 120, unless I start doing something—it may not even be practical . . . like making a list—it prevents me from focusing on those 'I just can't take this anymore' thoughts."

Similarly, just as I once did, many suicidal students will bury themselves in dull, routine academic busywork in the weeks before a suicide attempt, presumably to enter a sort of "emotional deadness," as Roy calls it, which is "an end in itself."

"For me," explained another man, "the most important thing when dealing with suicidal ideation is the escapism. Anything, literally anything, a book, music, going for a run . . . even doing something stupid like pacing back and forth or counting."

The grim, tedious details of organizing one's own suicide can also offer a welcome reprieve. People are often surprised to learn that it's not uncommon to find positive emotions in genuine suicide notes.* "When preparing for suicide," explains Roy, "one can finally cease to worry about the future, for one has effectively decided that there will be no future. The past, too, has ceased to matter, for it is nearly ended and will no longer cause grief, worry, or anxiety. And the imminence of death may help focus the mind on the immediate present."

Even professional therapists can be led astray by this positive change in emotion. In a study of clinical psychologists' experiences with losing a patient to suicide, the vast majority said that they'd failed to detect any danger just prior to the event. In fact, when asked to reflect on the days and weeks leading up to the suicide, these experts reported that, at the time, they'd have put these individuals at fairly low risk. "The patient showed signs of a brighter mood and was laughing with significant pleasure," said one practitioner.†

*In one old study on suicide-note content from the 1950s and published in the *American Journal of Psychiatry*, a team of researchers reported "a range of affect" in a large sample of notes, with positive feelings found in half, neutral feelings in 25 percent, and hostility in 25 percent.

†This was something that the sociologist Durkheim noticed as well in the late nineteenth century, observing in his book *Suicide* how such individuals "kill themselves with ironic tranquility and a matter-of-course mood."

Stage 6: Disinhibition

The poet Robert Lowell once remarked that if we came with an inborn kill switch embedded in our arms, so that by flicking it we'd die instantly, we'd all have flipped the switch long ago.

That's the reason I don't keep a gun in my house.

Consider how astonishing it is, really. Put a gun to your temple, and getting from *here*, where the tangible cosmos of your consciousness dwells, to *there*, where there's no more you at all, is but the span of a blade of budding grass, the stride of a broken toothpick. And with a copper-point bullet traveling at a velocity of 860 miles per hour tearing into an ape skull hopelessly forged by nature to withstand the occasional bumps and bangs, not incendiary high-caliber missiles, this road to your nonbeing is a frighteningly short one indeed. With all the debate over gun control, one seldom-discussed fact is that a considerable majority of gun-related deaths in the U.S. are suicides, not homicides. According to the Centers for Disease Control and Prevention, there were a total of 21,334 suicides by firearm in 2014, and 10,945 homicides. That's nearly *twice* as many gun-related suicides than murders. There are plenty of other ways to die too. But suicide by firearm may be the closest we'll ever get to Lowell's kill switch.

Normally, there are many things that help keep us plodding along. If you're religious, for instance, then thinking about the inherent "wrongness" of suicide can act as a moral deterrent (more on that in chapter 7), but of course there's also worrying about how our deaths would affect those we love, or even inspire copycats, and then, as we saw earlier, there's our self-preservation instinct under normal conditions of evolutionary homeostasis.

As a result of being in a cognitively deconstructed state, these barriers collapse one by one. The suicidal individual's capacity for meaningful thought is impaired; with a drone-like focus on concrete details, abstract thinking that would normally generate spiritual or other protective ideas about, say, finding hidden purpose in suffering are alarmingly absent. Shneidman once remarked that "the single most dangerous word in all of suicidology is the four-letter word *only*." Those who are intent on taking their lives, in other words, have entered a mode of dichotomous thinking characterized by all-or-nothing reasoning. The situation has become black-and-white; there's no metaphysical subtlety, only life-or-death.

When one has progressed this far along the steps of Roy's stage-theory model, they are in an altered state of consciousness far removed from normal experience. What about our bog-standard aversion to pain, though? Suicide hurts in more ways than one. And many methods that seem positively masochistic reveal just how shockingly tolerable physical pain can be when the only alternative is unabating mental anguish.*

Work by the psychiatrist Kimberly Van Orden and her colleagues sheds some light on the component of behavioral disinhibition. In addition to suicidal desire, the individual needs the "acquired capability for suicide," which involves both a lowered fear of death and increased physical pain tolerance. One acquires this capability by being exposed to related conditions that build up the individual's fear and pain-related tolerance. This is yet another reason why one of the best predictors of suicide is a nonlethal prior suicide attempt. "Suicide is like diving off a high board," writes Al Alvarez, "the first time is the worst."

A woman who'd halfheartedly tried to kill herself in a single-vehicle accident told my student Bonnie this about her second attempt:

*"To kill themselves," writes Kay Redfield Jamison in *Night Falls Fast*, "the suicidal have jumped into volcanoes; starved themselves to death; thrust rumps of turkeys down their throats; swallowed dynamite, hot coals, underwear, or bed clothing; strangled themselves with their own hair; used electric drills to bore holes into their brains; walked off into the snow with no provisions and little clothing; placed their necks in vices; arranged for their own decapitation; and injected into themselves every substance known to man, including air, peanut butter, poison, mercury, and mayonnaise. They have flown bombers into mountains, applied black widow spiders to their skin, drowned in vats of beer or vinegar, and suffocated themselves in their refrigerators or hope chests. One [patient] tried repeatedly to kill himself by drinking raw hydrochloric acid; he survived these attempts and died only after swallowing lighted firecrackers."

> That second time when I was going off the edge of the road I was like, "I really don't care. I hope I land hard because I don't want to land soft. I want to go *boom* and be dead!" Let the car blow up, whatever is going to happen, happen, and be completely done.

A history of other fear-inducing, painful experiences also places one at risk. Physical or sexual abuse, combat exposure, and spousal abuse can indirectly "prep" a person for the physical pain associated with suicide. This is one of the reasons why self-harm is so worrisome. And genetically heritable variants of impulsivity, fearlessness, and pain tolerance may help to explain why suicidality runs in families.

Van Orden and her coauthors even cite some intriguing evidence that habituation to painful or fear-inducing stimuli is not so much generalized to just any old suicide method, but often specific to the particular method used to end one's own life. For example, a study on suicides in the U.S. military branches found that guns were most frequently associated with army personnel suicides, hanging and knots for those in the navy, and falling from heights were more common for those in the air force. And it's not just prior exposure to painful stimuli that greases the wheels of suicidal behavior. Data also show that people in this final stage are more socially passive and submissive than they are normally, which may play a role in their *allowing* themselves to be subjected to physical pain. It's as if some people are genuflecting before us, saying: "I deserve to suffer in my death."

✷

For those of us who are inclined to find ourselves swept into these powerful waves of suicidal thinking, Roy's model shows us, essentially, how our minds can be tricked into making a fatal decision. In understanding what's happening to us, we're poised to outfox our own deviously self-destructive natures. We just need our second self, that wraith-like observer watching our step-by-step undoing, to keep us at an arm's length from our emotions.

"There's something oddly comforting about knowing how it

all works," I told Roy. "Of having a better understanding of what's happening to oneself. How do you see us applying your insights, in terms of prevention efforts?"

"I'd advise people to step back and don't get carried away," said Roy. "Know the process. See it from this other angle and evaluate whether this is a good idea or not. You can at least see it as a temporary state and say, 'Well I'll revisit it if I feel this way next month and then maybe I'll kill myself.' "

I know, I know. It's probably not the sort of inspiring message you'll find on a poster at the dentist's office. But what has that proverbial "Hang in There" kitten dangling from a tree limb ever done for you, anyway? For all we know, it was trying to kill itself.

Anyway, Roy's advice, gruff and straight to the point, may well save your life—or that of someone you love—one day. Let's worry about the day after that once we get there.

5

the things she told lorraine

It appears to me that we should never dispute the feelings of others; counsel can only operate on conduct, the laws of religion and virtue providing alike for all situations; but the causes of misery, and its intensity, vary equally with circumstances and individuals. We might as well attempt to count the waves of the sea, as to analyze the combinations of destiny and character.

Madame de Staël, *Reflections on Suicide* (1813)

On the fourteenth of April 2014, seventeen-year-old Victoria ("Vic") McLeod killed herself by jumping off a ten-story condominium building in Singapore. In her purse was a note: "If I'm brain damaged," it read, "I don't want to be kept alive. I don't want to be a vegetable."

Aside from that, she offered no explanation.

Seven months later, her parents discovered a journal on their daughter's laptop that she'd been secretly keeping in the four months leading up to her death. They shared this extraordinary material with me in the hope that her tragic story might help others. On reading it, I think Vic, too, would have wanted you to know her story. "I don't want other kids to feel like freaks," she writes, "or teachers to always assume that the kid at the back of class who never raises their hand is just 'shy' . . . when they are really paralyzed with [a] fear and hopelessness that they believe no one could ever understand."

Who was she? At the time of her death, Vic McLeod was just shedding the chrysalis of her youth to become an attractive tall and leggy blonde, hovering in that brief stage between gangly and statuesque. Uncomfortable in her own skin, she buried herself in books and writing. She was a gifted writer and poet, and, like many such souls, she was exquisitely sensitive. "Emotionally vulnerable," as she describes herself. "One word could tip me over."

A New Zealand citizen by birth who'd spent her whole life in busy urban Singapore, where she attended a competitive English-speaking high school for international students, she had aspirations of being a psychologist, could cite J. K. Rowling and Anne Sexton with equal ease, and always felt like an outsider. She was also, if one reads between the lines, coming to terms with unrequited romantic feelings for "Grace," an American girl whose family had moved to Singapore when she was in the ninth grade. Grace had since moved to a different high school, but the friendship continued—mostly online.

Vic's poignant and honest writing offers us a rare look into a privately despondent reality and, critically, it allows us to see Roy Baumeister's escape stages unfolding in real time as the deluding forces of suicidal thinking lay siege to a beautiful mind.

Similar to how many people begin their journal entries with "Dear Diary," Vic addressed hers to a fictional character named "Lorraine."

So, let us examine the things she told Lorraine.

Stage 1: Falling Short of Expectations

An only child with loving parents employed in the media industry, Vic came from a comfortable middle-class home where she wasn't really left wanting for anything. She even mulls over this apparent gap between her feelings and her outward situation in her journal. "I have so many opportunities," she writes. "If some people were me, they'd be so happy. They'd have their own room. They'd have a great school. They'd live everyday like it was heaven. Hence why am I being completely self-indulgent? I shouldn't be so wrapped up in my own stupid, worthless problems, when some people would give anything to live my life. I have my whole life ahead of me. So why can't I just live it?"

At the forefront of Vic's mind was a crippling worry about her impending school grades. She died on the first day of the new term, the same week that the results from her preliminary qualifying exams were due to be returned. "I've known that I will never have a dazzling life, what with the grades I get," she writes in her first diary entry. "But if I keep carrying on like this, I might actually end up snapping."

> I think (and I know it sounds melodramatic) that I might not make it this year. I know that when I see those grades bold and black on a piece of paper—I will either jump for joy, or jump off the top floor of this condo. I know it is absolutely ridiculous to kill yourself because you failed high school . . . but I don't know what to do.

Vic was struggling in a few classes and having difficulty facing the prospect of actually failing. "If I fail these exams, I know that my [grade point average] will be ruined and I can kiss University goodbye. I can kiss my life goodbye. All I want is to be a good psychologist. I can't do that without a proper degree *and* a Masters."

What makes Vic's story particularly frustrating to any objective reader of her journal, most of whom can easily sympathize with her teenage travails, is that she is actually keenly aware of the passing significance of her own problems—in the grand scheme of things. She is, in fact, remarkably insightful about her own state of mind. "I'm not that much of an idiot. There are worse things than failing school. But when there is nothing else worse than that in your life, it's the most terrible thing you can let yourself do."

Many students, of course, face serious academic challenges. Fortunately, most of them won't commit to ending their lives over the prospect of an F, no matter how scary that may be. But when a seemingly insurmountable life problem like this is compounded with other tempestuous individual traits—such as social anxiety, low self-esteem, and perfectionism—it makes for a dangerous set of conditions. "I will have failed my exams," writes Vic three weeks before she takes her own life. "And it'll all become a ticking time bomb. And I'd have to clean up the shrapnel. I will have failed, and

I have no other option. It's not the whole reason, though. If people thought that, I couldn't bear it."

A natural-born psychologist, Vic knows that her difficulties stem from a host of factors. Back in January, she describes a visit to her classroom by a rather jejune guest speaker. "He talked to us about idealistic things like 'motivation' and 'success.' "

> Oh the irony. He was a middle-aged guy in a lurid orange shirt and had a kitschy attitude. He told us how when we're sad, we're supposed to put a patch on it and move on. I wanted to build a case for myself to contradict a lot of things he said. He was simplistic. We all have our own ways to cope with things. We are not all factory-processed clones (although many adults would argue that). Our genetics are different, which would play a part in the fact that we all respond to adverse life events differently. Some of us have an increased risk of depression, which would thus affect our responses. I'm just saying he was very one-dimensional.

In the backdrop of Vic's school troubles was a brooding social life predominated by intense feelings for her friend Grace. Over the past year, Vic had developed an intimate emotional connection with this other girl. A perfect match for her own clever cynicism and self-denigrating wit, Grace is at the epicenter of much of Vic's thoughts, and she has become both a source of strength and despair. "I just want to say that it has been an honest comfort to have someone that understands," writes Vic in a letter meant for Grace. "Thank you so, so much for being there and saving my sanity. I just . . . you make me feel like I'm not alone, and that's a huge deal." The despair part comes in with Vic facing the dawning awareness that her romantic feelings for Grace are unlikely to be reciprocated. "I never really understood the value of honesty until recently," she reflects in her journal. "It can both heal and destroy you." Vic envisions a bleak future without Grace, and she grows convinced that the latter's romantic inaccessibility is a harbinger of a life of profound loneliness.

In a poem titled "Slim Hands," Vic captures this excruciating feeling of unrequited love in painfully vivid prose:

She laid her head on the pillow beside me,
Flyaway curls spread across the downy lint,
Her eyes closed, her spine convex,
Slim hands reaching for someone
In the empty space, though my warmth
Emanated, and my eyes breathed in
Every perfect lash, line and mark on
Her canvas that she did not paint, and which
She could not see, and every little piece of
Beauty she thought she could never be.
And though I was close enough to touch,
I knew those hands
Would never reach me.

On the verge of young adulthood, Vic also mourns the loss of her childhood, and with it, a certain standard of care she'd come to expect. "We dissolve into a sea of faceless, nameless humanity. Just another grownup with no one else to take care of you. When you're a kid, people genuinely care. You're young. People can't help but feel the need to help you achieve the things they couldn't. When you get older, people stop caring."

Stage 2: Attributions to Self

Throughout her journal, Vic frequently compares herself to idealized others, particularly other girls in her school who—in her eyes, at least—have everything going for them. She tells of being out with her friends and crossing paths with a girl from the neighborhood. "You know, one of those chicks that look like they have it all."

> Blonde. Lithe. Top grades. Popular. The whole jealously wrapped-up package. I mean she was *exercising,* for heaven's sake. Walking down Claymore Avenue with $200 Nikes and a cloned training buddy, no doubt to the gym. . . . It's kind of beyond me how anyone can have their life so sorted. Maybe I should start comparing [these girls] allegorically to filing cabinets. Each file section is a subdivision

> of life. Academics. Family ties. Extra-curricular activities. Social stature. Looks. Boyfriends/girlfriends. Socioeconomic state. Mental health. Physical form. . . . Not only is every section perfectly organized, but also each page has the right border, font, page numbers and grammar with A-plusses on each sheet of crisp white paper inside every pastel folder. And if one of those papers gets dented for even a second, there's some kind of top-of-the-line printer hidden in there that reprints it, crumples the old one, and throws it away. . . . Sometimes my folders manage to get printed. But then the printer breaks down. And the old copies pile up and up. Until there will eventually come a point when the cabinet won't be able to fit it all in. This is known as "cracking." I gave up on trying to be [that type of girl] long ago.

Vic has little negative to say about other people, but she's almost unrelenting in her self-disdain and in her certainty about never being able to overcome what she perceives as her shortcomings. "I am the definition of a hypocrite. An angst-ridden, over-sensitive delinquent . . . I'm a stupid fucking drama queen."

"It's not other people. It's just me."

In another entry, she admonishes herself for still being in her pajamas at three in the afternoon. "I'd better give myself instructions," she writes with bitter sarcasm. "*You have to get up, Vic. Now you have to get dressed. Wear something discreet. Now you have to fold the clothes from the dryer*. Maybe I'd better take it slow. *Pathetic. Just pathetic.*"

Vic describes waking up in a panic, realizing how far behind she's fallen in her courses. "I woke up early with my heart pounding,"

> . . . it was seven in the morning and somehow the cortisol or adrenaline or whatever chemical it is forced me out of bed. I was thinking about all the untouched homework. Kind of shocked me how much I hadn't done and how much I was supposed to do. I tried going back to sleep but I just couldn't stay still. I kept thrashing around like I was having a nightmare. It was that stage between a panic attack and that nauseous fear clawing at your stomach. . . . Dammit I wish I would stop feeling sick. Jeez, why do I always do this? I'm an idiot! I had

> the whole fucking holidays! I have so much work to do, it's insane. And the year has barely begun.

As the journal progresses and Vic continues to have suicidal thoughts, she reflects further on her relationship with Grace, convincing herself that she's become a burden to her friend. "I'm not worth her time. I'm not worth anyone's time. I'm toxic to her. She doesn't need to be held back by me. She's got her whole life ahead of her. She's so much smarter and kinder and funnier than she thinks. . . . She'll be okay. They'll all be."

In an entry written two weeks before her death, Vic's self-loathing reaches a crescendo. "I don't deserve to be alive. I don't want to face the trials of reality, which is obviously cowardice. I have let this cowardice envelop me, and I can't shake it off."

> . . . I don't want to fight. I don't want to live. I have had nothing bad happen to me except my own doing. I don't deserve happiness. I don't deserve to complain. Take it from me and give it to someone who needs it. Let another complain because they deserve to. Please just get rid of me.

Stage 3: High Self-Awareness

To say that Vic expresses intense self-awareness would be a gross understatement. She's not merely "self-absorbed," as a narcissist would be. Neither is she simply "self-centered," as she describes herself. Instead, as part of the suicidal process, she finds herself locked into a rigidly myopic gaze, one in which she cannot peel her inner eye away from the source of her anxieties. She writes of how other people, by contrast, seem "disconnected from their own darkness" and seem easily distracted from any problems they may have:

> It's kind of surprising how superficial everything is. You talk about TV shows and music and grades and all the same stuff, it makes you think people are disconnected from their own darkness. Or perhaps they just don't allow themselves to become sad. They put a patch on

> whatever wounds life throws at them, bury their feelings, walk it off, forget about it and move on.

The thin veil of artificial meaning has been lifted for Vic and, as a consequence, she is increasingly desperate to escape from her own consciousness, which has become punishing. Whenever her thoughts wander fleetingly to more pleasant things, she's unable to keep them there for long. "I'm still marveling at the fact you can thrive in a world of inner happiness," she observes. "And then let peripheral reminders crash through and rewire your brain to make it think that pursuing contentment is useless."

Vic's social anxiety, which she identifies as her core problem, gets worse. She describes her fear of being asked to do physical exercises in front of her classmates during her health class. "I hate the subject because I don't know anyone in it."

> And there's always the terrifying possibility that we could have to do a practical, where I get to show my un-sportiness and unpopularity, because the class is small and the only people who are in it are popular high-achievers. If I got 90s in my other subjects, I'd drop Health like a shot. But I'm surprisingly okay at it and it's an interesting subject. But the mere fact that the class is composed entirely of people who go running every day for the hell of it and fit in our cohort so well makes me paranoid that they think I'm the worst in the class and an avoidant personality freak.*

*The social psychologist Tom Gilovich did one of my favorite studies ever. Gilovich had undergrad students change into a colorful T-shirt featuring 1970s crooner Barry Manilow, then sent them into a room to mingle with other students. I happen to like Barry Manilow, but when this study was conducted in 2000, it's safe to say that he hadn't exactly been on the cool list for a while, and so your typical eighteen- to twenty-one-year-old would have found this to be a rather embarrassing experience. After the gathering, Gilovich and his colleagues asked the students to estimate the percentage of people in the crowded room who'd noticed their shirt, comparing this figure with the *actual* number of people who said they'd seen the Manilow attire. On average, the red-faced students were convinced that at least half of the room must have seen their fashion faux pas, but in reality the T-shirt had registered in the minds of only a small fraction. Gilovich calls this universal psychological tendency to think

Feeling increasingly cut off from her parents, Vic believes that any efforts at communicating her distress to them would be fruitless, so inaccessible are the recesses where her suffering lies. She refers to the father of Jewish Holocaust victim Anne Frank, and how he had no idea of the true depth of his daughter's spirit as it's preserved in her famous diary.

> It's quite incredible how much parents don't know about their kids. They honestly have no fucking idea. They don't get it. They just don't. I think Otto Frank was right. Parents will never really know their children. And it's always the worst things. It's just incredible how a parent can have no idea that their son or daughter wanted to kill themselves. How many times they were oblivious to the moments where they could have lost their child. Or the times when their kid would have done things that they could never have understood, like slicing the skin from their forearms. Or simply the impossibility of not caring what others thought of them in a classroom. Or anything else that consumes the hidden parts of their mind. There will always be the bad, ugly components that parents will never understand. That's perhaps why we don't want them to know. We know that they won't get it, even in the rare circumstances that they try.

The one thing sustaining Vic in its ability to dilute the potency of her self-awareness is speaking with Grace, whose understanding is an anodyne and greatly comforts her. "Seriously, I never imagined I would say any of those things to anyone. Let alone someone who *got it*. . . . Thank Christ for honesty. She gets it all. Thank. Freaking. Christ. . . . We talked for ages about stuff we hadn't said before. I would never ever in a million years have considered that she got it about suicide and the whole self-harm thing."

---

that we're at the center of other people's attention, when in reality we're probably more on the periphery, the "spotlight effect." Teenagers, in particular, are notoriously prone to suffering this unnerving illusion of an ever-attentive imaginary audience, an aspect of human development that can make adolescence—when everything that can go wrong with one's physical appearance has a way of doing so—an especially painful part of the life span.

> . . . I spoke of things I would never speak to my parents. Now a once-stranger with not a drop of my blood in her veins knows me better than one whose eyes are almost the same as mine and whose combinations of DNA formed my very existence. I almost laugh at this absurdity. How can one who has no relation understand me as if a part of their very being is shared with me? Life is an inexplicable thing, for those who prophesize too much.

The theme of escape is prominent throughout Vic's journal, and she's shrewdly knowledgeable about the range and function of psychological tactics that give people a much-needed distance from their thoughts. She fantasizes about travel—seeking a literal escape from her troubles by leaving everything in a trail of dust. "I want to go on a train."

> There's something numbing about going on a train. I don't quite know what it is. It gives the illusion of taking you far away, taking you away from everything. It's not like a library, where you have to look like you're doing something, and it's so quiet, you can hear other people breathing. On trains, the whirring of the tracks, the constant sounds as it reverberates down a dark tunnel with no end. It just keeps going and going and you never have to worry about it ending. It just keeps going. You don't have to do a single thing on a train except travel. When I fail school, I might just spend hours on trains. It's not a very normal thing to do. I don't know anyone who would substitute train travelling for [escape mechanisms] like booze or cigarettes. But I just want to do that.

As an alternative to this endless train journey to nowhere, Vic tries to employ more practical means of escape. She listens to music. "I listen to sad descants and instrumentals." She reads. "I read Anne Sexton, a poet whose words uncover beautiful and terrible reality." She tries to sleep. "Right now I'd be happy if someone flicked a switch that would make me sleep for a long time." She writes in her journal. "I think [writing] calms me down when my thoughts get out of control. But on the other hand, it can also fuel them. Words, I mean."

In the following remarkable poem, titled "Good Girl," Vic's lucid understanding of how different people use varied means to escape from a debilitating state of high self-awareness is even more clear.

I snuck a lighter from the kitchen
When I was fifteen; not to smoke: to smell.
I lit incense stick after incense stick,
And inhaled sandalwood smoldering,
As the drunkards climbed the stairs,
Swallowing tar and solute methane,
But really, what is the difference?
We escape our minds with our senses:
Chemical compounds to rid
Us of our demons.
They laughed the night away,
And I heard their chortles echo
As they strolled the empty streets
To rid themselves of the pain
Of guilt and all the trials
Life so unpityingly hands us,
And to endorse the transience
Of which we make
By a swig, a drag, or a cut,
A smell, a kiss, or
A word.

Stage 4: Negative Affect

Vic is in psychological distress, but with the exception of her heart-to-hearts with Grace, she keeps her pain mostly to herself. In a counseling session, she decides to open up about her suffering with her counselor ("Miranda"), but she only reveals so much. There's no mention of her suicidal impulses, which are growing in frequency, yet she discloses her cutting behavior—a clear response to the mental pain she's experiencing.

"Well, Miranda knows I self-harmed. I'd never have considered she'd work that out. Thank God she hasn't told [my parents]." Vic describes for the counselor a recent evening in which she'd felt particularly desperate. "I explained that it was like those thoughts of hopelessness had gotten out of control and I wanted escape. She thought I meant by the blade. I was talking about death. Though I tried the blade. I didn't cut deep enough."

Vic's journal is punctuated by moments of graphic despair. Her anguish is palpable. Suicidal thoughts begin to encroach on her more regularly as the calendar marches on. "Today was bad," she writes a month before she dies. "Sat in the shower. Did the whole crying bit. Sat in bed. Did the whole sad songs and crying bit. I thought about death too many times today. Wanted to be by myself again at lunch. Stuff to tell Miranda: wanted to be by myself today, incessant suicidal ideation, just fix me. Please fix me or I can't be here. PLEASE MAKE THIS SAD STOP. FUCKING MAKE IT STOP. God, something out there, please make it stop. Just make it stop."*

Vic notes that she hasn't felt happy for weeks. A "sedating melancholy" has taken hold. She notices that her ability to contemplate a more positive future is compromised by her negative affect. "It's a funny thing about emotions. When you are happy, you know that it will dissipate, but in that moment it feels like it could last forever. When you are sad, it is long. A protracted and dull lethargy that is dense, and you forget that it won't last forever. Darkness lures you in. You almost welcome it after telling yourself not to."

Yet in this emotional state, she succumbs to pessimistic reasoning and mistakes it for rationality. "I know realistically that this anxiety thing is not going to go away. I will live my life completely alone."

Even when she allows herself the possibility that she will finish high school and be accepted at the University of Otago in New Zealand, where she plans to apply, the future looks bleak. "Everyone will drift apart. I will be alone in a small town where I don't fit in. I'm not a New Zealander. I don't care that it's on my freaking passport, I am just not one. I don't know how to adapt. I would miss Singapore too

*There's no indication that Vic ever actually shared her suicidal feelings with Miranda or any other adult.

much. But at the same time, I can't stay here. I don't think I'd ever really find anywhere."

Vic comes to expect a permanent imbalance between happiness and misery, with the latter being the baseline emotion that will color her entire life. A few weeks before she jumps to her death, she notes how—to her surprise—her mood briefly lifted. But she brushes this transitory relief away as a sort of seductive ruse, an illusion that she's seen through. "I got distracted by fleeting moments of egocentric happiness a few days ago. I momentarily forgot that happiness is just a distraction from the reality of life."

In a poem titled "Winter Mind," Vic paints a mournful image of her ongoing sadness.

It's a shadow that sweeps over you.
Grey skies. Charcoal streets coated with grime
And glistening wetness under rusted lampposts.
Rain-splattered glass, the drops sliding down car windows like tears.
The eternal, slow, steady rain falling, on cold days through fog.
Tangled, windswept hair. Stale black coffee.
Dust building on the spines of unopened books.
The color draining from floral-patterned sheets and curios
In bedrooms like tea-dregs, giving way to crumpled black clothes
And silence.
Splinters of dark aluminum sunrays refracting
Through the gaps in drawn curtains.
Time passing in a colorless blur between day and night.
The weeks stretching into an endless winter.
Kept inside to rot like a dying plant. Eyes opening only to wish,
To cling on to the only thing you know; that death will come to you.
Sad songs become a daily commodity.
Not even talking can expunge this melancholy,
There will soon be nothing left to extract.
Silence.
The time has passed for brutal honesty,
For there is no energy left to bellow out the troubles of your world.
You want to be alone, but you crave company.

But in your heart you know that no one wants to be dragged
Into your toxic, morose-ridden domain.
It's an emptiness that food cannot fill.
A detachment that paralyzes your neurotransmitters.
Cold numbness.
Wait for the high, it'll come some time.
For now, there is nothing but your thoughts,
And silence.

Vic is frightened.

"If I don't do something, it's all going to spiral out of control," she writes.

Stage 5: Cognitive Deconstruction

As her suicidal feelings intensify, Vic finds herself in a state of ennui—that stifling gray zone in which passions dull, plans idle, and time slows to a crawl. She seems to have a vague sense that something has changed. "It's what you might call a non-cataclysmically-but-still-destructive coping mechanism," she observes of this shift in her thought processes. "I'm neither hungry nor full. I'm neither bored nor active."

Yet she can't snap out of it. "God knows I should be running around using up adenosine triphosphate stores and diligently studying like an actively functioning young adult. But here I am, sitting here. Wasting precious time."

The stress of her looming assignments immobilizes her; she knows what she *should* be doing, but knowing does not lead to feeling. "I'm tired of this. They're putting too much pressure on us. I'm tired. I told myself to get up and do my homework like a good girl. But I'm just sitting here, way too tired than I should be."

More worryingly, Vic notices that her ability to envision the future is being compromised. "Right now the future seems like a dark, obscure place that I can't see myself in. There's nothing ahead. Nothing."

Feeling hopeless, she begins to settle into numb indifference.

"Like now I'm thinking 'why bother?' Why even bother if I'm setting myself up for failure."

> This kind of thinking is more unhelpful than usual, because this is the point where a lot of people stop trying. And the whole thing kind of intensifies into this huge downward spiral, where shit actually happens and people begin to screw their lives up. I had all these plans. Now I don't care. I suddenly don't care about my future. I don't care if I disappoint my parents as long as they don't have to face me. I don't care if I do nothing and just sit in a room wasting my life away while everyone else moves on. I don't fucking care if I can't afford anything. Sure, I'll be filled with regret for the rest of my life, but I'll be alone. I could just run away from everything.

In a short poem titled "Cavity," Vic describes this in-between feeling, of being stuck in an unhappy present where the past is gone and the future is for others, in uncanny detail:

> *She is empty.*
> *She is the untenanted space*
> *Between the train and the tracks,*
> *The filter between the mind and the face,*
> *She is the void between the glass and the frame,*
> *A speck of dust on stained quartz crystal,*
> *And the moon's veil between wax and wane.*

Time hurts when nothing happens. The worrisome future is held at bay by an enflamed present. In another brief poem, this one untitled, Vic shows us how an ominous disquiet lurks there:

> *She prayed and prayed,*
> *The covers wrapped around her,*
> *Listening to the clock tick away her time,*
> *Until the day she left the bed,*
> *And left herself behind.*

The urge to escape is still strong in Vic's writing, but she's increasingly unable to muster her trusted psychological diversions. "For the past three days . . . I did not think thoughts that I now want. Such as doing sedating things that do not make sense, like catching a train and traveling without a destination. Just traveling. Seeing the people as nothing more than chromatic hardcovers with untold stories. Replications of dyed hair and dark eyes wanting and waiting for approval."

Instead, she allows her febrile thoughts to boil. "I'm sleeping late tonight. I'm letting the triggers engulf me and overpower the [mental] vacation, with the gritty truths of the disillusionment that is life."

Soon, she finds an effective means of escaping the suffocating present—but it's a terrible one. Vic begins to busy herself with details of her death. She writes a poem, "I Will Be," which she hopes will be inscribed on her tombstone:

You look up.
Stars
—little lights,
Little places where you
Cannot live with your
Lungs and your skin—
But with your heart.
My little world will meet
Yours, and yours mine.
My organs, though left
Behind in the earth to become
The trees, or to live and
Be wanted in another's body,
I will be there as
Another dot to join the dots,
So that when you look up,
I will be there, and you know
I will be free.

By this stage, Vic's mind seems tragically made up. "If I'm getting my 'affairs' in order, I might as well sort some things out. Here are my things to sort out":

> . . . my heart, lungs and kidneys will go to someone who deserves them. The rest of me will become nutrients in the earth, like how Sylvia Plath wanted to be recognized by nature, adopt its immortal and daring qualities, and be useful in a way she thought she couldn't be in life, in her poem "I Am Vertical." That the song "Lying to You" be heard. Grace told me about that song. It's awfully fitting. It's beautiful. My clothes and shoes (or whatever wearable ones) go directly to people who could never afford them. My books too. My little silver crescent-shaped moon necklace that I have worn every day be given to Grace, if she wanted. The necklace is very special to me. I bought it so that I could have a little faith. She is a pragmatic scientist by nature, but I just think she'll need it. The cash from my college fund that my parents spent years saving up, be spent on a nice trip. A break they deserved to have a while ago. Or else be given to a charity. One that raises awareness about social anxiety, because God, it needs to be. . . . My mediocre poem "I Will Be" be put on my headstone.

Then, believing that her journal will never be seen, Vic laments the unlikeliness of these outcomes ever materializing. "But no one will know about this."

> I'll probably end up getting buried with a garish headstone with the whole "Beloved Daughter" et cetera, my heart, lungs and kidney still in my body, my little necklace thrown away. My books collecting dust in old boxes. My clothes in the garbage. But in the end, it doesn't really matter.

Finding solace in the promise of no tomorrow, Vic's mood stabilizes. Her anxieties ease. But this is a dire sign. She mentions an argument she has with her parents over her poor grades. "I strangely didn't get overwhelmed like I usually would when this happens. Not sure whether I should be relieved or alarmed."

In the following untitled poem, Vic exhibits the emotional leveling effect that so often characterizes a suicidal resolve:

Red eyes with tears, fossilized,
I am almost peaceful.
There is a strange acceptance about expiry,
Finality.
Closure.
Nothing more can taint your head.
No more voices can eat away
The gossamer strands of thought
Like silkworms devouring matter,
Piece by piece,
Until there is nothing left
But an empty gap.
The worms are hungry.
You provide no nourishment.
No seeds planted, waiting to grow
And bloom with motivation
That serves as a well.
It has been drained through
Your eyes. Perpetually saturated.
You're as helpless as a rotting leaf.
The wind comes.
You blow away.

Throughout her journal, Vic expresses strong reservations about killing herself over the pain she knows her death will inflict on others. "The only thing stopping me is them," she writes of her parents. Speaking of Grace, she adds: "I can't fucking do it to her right now."

Slowly, however, an invasive apathy eclipses any such feelings that might keep her from going through with her plans. The ennui has worn away Vic's capacity to care. "At this point, I don't care that I am being selfish. I've started using suicide as my comfort blanket. [My parents] don't know that I won't be here by the end of the year. It'll kill them. But I have to stop caring. It's my escape, something I will always have access to."

There's evidence of a battle between her concern for others and her striving for oblivion. "I taught two people the value of life," she writes of her parents, who have no other children. "You will lose me. My god, it will destroy you. What I will do, will be unforgiveable." Intellectually, Vic can't bear what she's contemplating doing to the people she loves, and who love her. Yet she's also being held hostage by emotions that are interfering with her compassionate and considerate nature. "I will commit the worst thing you can ever do to someone who loves you: killing yourself. The scary thing is, I'm okay with that." Note the expression *the scary thing is*; it tells us that Vic is cognizant of the fact that such feelings aren't like her . . . that is to say, a part of her—the healthy part with infinite potential—knows that she's caught in a process that is beyond her control and is helplessly watching the tragedy unfold. It's as though she's written herself into a script.

"I am not sad so much as gone," she writes.

Stage 6: Disinhibition

One evening, two months before she took her own life in the very same manner, Vic finds herself on the top floor of an apartment building, peering down at the empty parking lot below. "I knew I couldn't go through with it. But I left the house and walked to the other apartment block that had ten floors. I got the lift to the top and wanted to just look down and see how high it was. . . . Suddenly, I just . . . wanted to be dead."

She describes this moment as the culmination of a "lingering despondency" over her "failure at not doing [home]work sooner, convinced that I was now a total failure, total lack of motivation, fears about the future." By standing there and imagining leaning into the wind, Vic is essentially inoculating herself against the natural fear of falling from a great height, a normal terror that could otherwise preserve her life. She's grooming herself, whether deliberately or not, for the terrible act to come.

The presence of people suddenly disarms her that night. "Two [families'] front doors were open. To let in the top-floor breeze into their living rooms."

> I got scared. God forbid, they'd see a white girl looking down over the edge. So I got the lift back down. I wanted to be isolated. To find a place where no one could find me. I saw the old fire exit staircase. There was an alcove underneath the narrow stairs, enough to fit two or three people. It was dusty, but well concealed. I stooped, sat there, and stared at the wall. At first, I was afraid someone would find me, but I heard no one. I listened to sad music on my iPod and just sat there, feeling numb and hopeless at the same time. I was both passive and scarily pessimistic. I wanted to stay there forever. I sat for an hour and I cried very quietly in a deadened sort of way.

In earlier journal entries, Vic expresses a characteristic ambivalence about suicide. On January 17, she writes, "I need to stop thinking about death and suicide and all this crap."

> I still want to be dead. But I want to get better. I need this to stop. But I can't do that to my parents . . . it would kill them. How do you go up to someone and say "I want to stop wanting to kill myself, but at the same time I really want to." And that you want to stop the paranoia. The sadness. All the shit that goes on. How the fuck can I do this?*

*What's especially heartbreaking are the many messages of support that Vic, despite her own private suffering, was sending to Grace during this time. "If you ever feel sad or lonely," Vic wrote in a letter to Grace, "please remember that you are a living, breathing, intricate, strong, independent [person] with the ability to love and to laugh and to cry. You have empathy, intelligence and kindness, and Grace, you're pretty damn awesome and if other people can't see it, they are deluded. And I should confirm that you are a whole lot prettier than you think you are. . . . I would sometimes think of the old, slightly clichéd (but nonetheless true) phrase 'this too shall pass.' Because no matter what happens, the bad things will pass. It's comforting to remind yourself that whatever future stress you'll endure, it'll be temporary—clear skies and Mary Poppins and all that are on the horizon! And even though good things pass too, don't be upset because it's over, smile because it happened . . . so Hakuna matata and *carpe diem*!" (It was this sadly ironic material that made Vic see herself as a "hypocrite" for not heeding her own mental health advice.)

By mid-March, however, Vic's thoughts have turned resolutely toward death. "We are each given a life. We're supposed to live it. I don't. It's as simple as that." She follows this decisive pronouncement with a strikingly elegant description of the kind of dichotomous reasoning that's so idiosyncratic to the final stage of Roy Baumeister's model.

"It comes suddenly and then dissolves," she writes of this constricted thought process.

> It chokes the fibers of your trachea and engulfs the acid in your stomach, releasing the poison into your lungs. For a second, you can't breathe. You forget how to breathe. You forget all those hopeless precepts, which are suddenly trapped in books of childish fairy tales. Reality stares you hard in the face. And reminds you, yet again, that you can't do this. This time is different. This time, every part of you, every fragment of your being, knows that it's right. The endless road to somewhere is blocked. You can't go forward. You can't go back. *You only have two choices.** You can let the rain and wind pummel onto you and bury your bones into the sediment that is under your feet, forcing you to wait a tormented, hopeless lifetime in that very spot until you are so far below the earth, that it sucks the last breath out of you, and carries it in the wind to dissolve in the sea. Or you can summon the coward's strength to ask the lightning to throw one last strike onto you, thus removing your selfish misery and despair for good. . . . But as you contemplate this, you begin to feel the earth swallowing your feet like quicksand. The poison that choked your throat subsides. The pounding rain and thunder that cracks across the sky recedes as time moves forward, and you stay rooted to the spot behind the towering wall. And then it comes. You cradle yourself into it; although whatever trace of light that is left is telling you not to, you reluctantly accept the numbness in your hands like an indispensable gift. It is in this moment that you truly know you have given up.

Shneidman's linguistic red flag—that four-letter word *only*—slips

*Italics added.

more and more into Vic's writing. She's seeing only two options now: live a life of unbearable mental suffering, or put herself out of her misery.

The disinhibition process is under way. "I'm scared," she writes.

> I'm really, really scared. I have to do this. . . . I'm just so scared of it not working. It has to work. God, it has to. It's now stuck in my head. I will think about it every second for the next few weeks. My heart feels like it's beating a little faster than it should. I get restless. Then I get numb. I just want to end this. It feels different this time. . . . It's all I have been thinking about. I know the details. Now there is finality. I have failed and this is the only option I have.

Her thoughts are completely consumed by suicide. "I keep looking up at buildings, thinking 'that one would be high enough' . . . it's like the idea has stuck."

> I know when and how I am going to do it. I can't back out of it. It isn't an option anymore. I have to fight the instinctive urge to stay alive. It's going to be the scariest thing I have ever done . . . I can't stop thinking that it would be better if I were gone. It's like a seed that's been implanted in my brain, and now I can't get rid of it. I keep visualizing myself standing over the edge. But it's even scarier to think [of] not doing it. When I was fourteen, going to the top of a ten-story building, looking down, convinced it wasn't high enough, and letting the fear engulf me enough to scare me out of doing it. I can't let that happen this time. . . . I need the shit in my head to stop. I have to do this.

Toward the end of her journal, Vic continues to weigh the emotional toll that her death will have on those she cares about, but her desire for a permanent escape has overcome her. "I will be that girl who is talked about. Whispers about her little incident."

> I will make a girl think she failed a friendship. I will make two people think they failed at being parents. This will be the last time I feel

the fear, and do it anyway. I will do it. I have to. I have to go. I will be that girl who was sick. Sick in the head. I don't think I am. I just want to go.

✷

When Linda McLeod went to get her daughter up for school that Monday morning, the first day of the new term, it was about 6:45 a.m. Still half-asleep herself, she turned on the lights of Vic's room only to find the girl's bed empty and the covers thrown aside.

Vic always made her bed.

"At first I thought she was having a game," Linda told me. "I started to look everywhere, and then, you know, when I couldn't find her, I just felt sick. I woke Malcolm [Linda's husband and Vic's father], and we went around looking for her and texting her friends."

A long-standing copy editor at the *Straits Times*, Singapore's leading English-speaking newspaper, where Malcolm also works as deputy picture editor, Linda is a keenly intelligent woman in her mid-fifties who gives off the careful air of an investigative journalist. But the story that haunts her the most these days, needless to say, is her own—how she and Malcolm lost their daughter to suicide and then discovered her startling journal seven months later when they got her laptop back from the police. Clearly titled "Vic's Diary" as a folder on her desktop, it was as though she'd left it for them to find.

But that day, that horrible day, they knew nothing about the journal, about the things Vic told "Lorraine."

It was the condominium's security guard, a kind man of Indian descent who'd known their daughter since she was a little girl, who came up to the couple and, through choking sobs, told them that something was seriously wrong. "He just said, 'Come, come.' He was absolutely broken. 'What is it? What happened?' we asked him. But he couldn't get it out. He beckoned us to follow him, then he flagged down a stranger in their car to drive us to Vic."

Her body had been found in the parking lot of a nearby apartment block.

"We were taken there," Linda said. "And there she was, and there were all these people around—lots of people taking photos—and crime scene tape."

A taxi driver had been the one to find Vic's body and call it in to the police. Because the inspector hadn't yet arrived when Linda and Malcolm got to the scene, the distraught parents weren't able to touch their daughter's body. Instead, they were made to sit beside her in portable chairs. Vic had landed face up. "Her face was preserved," said Linda. "It was undamaged. . . . I thought she looked regretful, but Malcolm thought she looked peaceful."

Before she'd jumped, Vic had texted a brief good-bye to her friends,* set her phone down, removed her flip-flops, and arranged them neatly at the edge. A long scrape mark on the side of the building, which Linda only observed weeks later, after the initial shock had subsided, suggested to her that Vic had adjusted herself as she fell, landing perfectly on a very small tiled area. "She was so focused even when she jumped," said Linda. "Any deviation, and she would have crashed onto the parked cars."

There's no way to know for certain when Vic had snuck silently out of her family's apartment early Monday morning, but her parents believe she jumped around 4.30 a.m. "Later, I made enquiries, left messages," Linda told me. "And eventually I found a maid who lived on one of the floors below. She heard a *whoosh* and then a thump around that time. She was very scared to talk, thinking she might be in trouble because she hadn't notified anyone."

Linda described the emotional scene on the ground that awful morning . . . sitting on the chairs next to Vic's lifeless body, waiting for the inspector. "Malcolm was . . . well, I think he nearly went mad," Linda said. "He pulled away and he got to her and kissed her before he was dragged off." She paused to wipe away the tears. "You know, it's just the kind of stuff that really eats you up for the rest of your life."

"I can only imagine," I said.

And it's true. I can't possibly know her parents' immense sorrow.

*Linda has never personally seen this group text, but she was told by one of Vic's friends that it was something along the lines of "Love you all, sorry guys."

Yet I felt like I'd come to know Vic through her crystalline words, and hearing from Linda what happened once she closed her journal that last time was hard for me. I don't have children myself and probably never will, but the truth is, Vic had managed to stir in me something rather surprising, what I suspect are dormant parental emotions. I found myself wanting to reach out through space and time and put my own clumsy gay wing around her, to help usher her through this thorny psychological world and search, side by side, for beauty in the absurd.

Had things turned out differently, I told Linda, Vic and I might have crossed paths.*

"Perhaps," said Linda, a bit wistfully. "The what-ifs are painful."

For her and Malcolm, there are plenty of those to contend with. In psychological parlance, it's known as *counterfactual thinking*. "Who among us has never wondered about what might have been had some past choice been different?" wrote the cognitive theorists Kai Epstude and Neal Roese in a 2008 issue of the journal *Personality and Social Psychology Review*. "Counterfactual thoughts are mental representations of alternatives to past events, actions or states. . . . [T]hey are epitomized by the phrase 'what might have been,' which implicates a juxtaposition of an imagined versus factual state of affairs." It's the stuff of regret, in other words, and for suicides, the counterfactuals can be an especially bitter yield. What may seem torturously clear to us in hindsight was anything but when seen through the lenses of one's limited knowledge at the time.

"What you have to understand," Vic wrote in her journal, seemingly anticipating that naive question being asked of her parents—*How could they not have known?*—"is that a lot of people hide it remarkably well. We build walls around us with smiles, laughs or silence. Silence is so easily mistaken for your classic case of 'moodiness.' Or the excuse that 'she's like that.' "

I asked Linda about the final few days of Vic's life. It was the tail end of the mid-semester break, the last weekend before school was

*I took up my current position at the University of Otago in July 2014, two months after Vic's suicide.

to start back up again and those dreaded grades were due to be returned. ("I know I will do it," Vic had written a few weeks before. "If not the last day of the holidays when Grace comes back [from her trip to the United States], then the same week when I receive my grades with the big, red low percentages.") Wanting to hang out with her daughter before things got busy again, Linda had taken Vic to get a pedicure with her that last weekend. "Normally she liked that sort of thing," Linda said. "But she was really unsettled in that chair. And the color she chose for her toenails . . . it wasn't a trendy color. It was a morbid blue. Later, when she was dead, and you saw the same color on her nails, you realized that was a reflection of her inner self."

"So, I guess that was a deliberate choice?" I asked Linda. "That she knew that would be the color she would die in?"

"Yes. Also, earlier that week, it was my birthday. And she gave me—I still have the card here, I think . . ." Linda rummaged through some papers on a shelf and retrieved the greeting card to show me. It had an unusual cover, with flowers set against a solid black background. "It's not exactly a happy birthday card," Linda said. "Because of the black. It has the effect of a funeral wreath, actually."

She lowered her glasses, returning to the shelf to look for something else. She came back with a paperback book this time. "And she gave me this book, a stupid book, called *Cleo*, by a New Zealand author called Helen Brown. It's about a mother who finds some sort of emotional satisfaction after her child is killed by adopting a cat."

These cryptic clues to her suicidal intentions are difficult enough for Vic's parents to bear, but still other events transpiring that final weekend show why the "counterfactuals" will always plague Linda and Malcolm's thoughts.

"The real thing that gets me," Linda said, "is that she was working on an overdue visual arts project. The teacher had been at me that Victoria was procrastinating and hadn't gotten enough footage for a video assignment. So, I said to Vic, 'You've really got to get this stuff done before school.' That Saturday and Sunday, then, the last days of her life, a lot of it was spent filming. When we look at the video now, it's awful to see . . . I can't believe we didn't realize . . . because she showed part of it to us on that Saturday afternoon. [In the film], she'd gone up to where she'd actually jumped from—but of course we didn't know then—and she got some tissue paper and had written on it 'Fear of Heights.' She'd set the camera up on the ledge and she shows herself bunching that tissue paper together and blowing it off the edge. I mean, how could we miss that? But because it was beautifully shot, Malcolm and I just said, 'Wow, this is so artistic, this is so conceptual' . . . you know?"

Still, something about the footage bothered Linda. "After we saw it," Linda told me, "I knocked on her door and said, 'You'd never think about doing something like that, would you? What you showed in the film?' She was sitting on her bed. She looked at me and said, 'No, Mum, of course not. Don't be ridiculous.' " Only later, of course, did Linda recognize the brief hesitation and the flash in Vic's eyes at being found out.

Like many suicidal people, Vic displayed an ominous sense of ease—an acquiescence to her enormous decision—in the hours leading up to her death. "During that last day," Linda said, "she was calm. Unusually calm. We went to the mall and she bought me this special pancake treat that I like, she returned a library book, and the last meal we had together, she helped Malcolm make spaghetti Bolognese. We had dinner watching a TV show about cooking that we loved and made a few jokes about the program. . . . Vic even imitated the presenter's awful accent."

In retrospect, Linda can see how calculated Vic's actions were that Sunday. Even the small things they talked about, what seemed at the time no more than insignificant musings, were, in fact, ruthlessly broached items. "There was a tidying up of things she wanted me to know," Linda said. "At one point she said to me, 'Mum, I want you to know, that you never blend your concealer properly, you've got to blend your concealer,' and she got it and rubbed it very gently on my face and showed me how," recalled Linda. "Then she said, 'There are some other things you need to know' . . . and I remember thinking to myself, 'I wonder what she's talking about.' She said, 'First of all, you know I've got no fear of heights, right?' And I said, 'Oh, what do you mean?' And she said, 'Remember that time Dad and I did that bungee jumping?' I said, 'Yeah, you were so brave, it meant nothing to you.' And she said, 'Yeah, and Dad couldn't jump properly, he was so scared, wasn't he, he'll never jump properly.' And then she goes, 'You know that wasn't the only time I jumped . . . when I went on that school trip to Malaysia,' she says, 'there was a sailing ship and all the other kids jumped over the side, but I went up to the mast, and they didn't think I could do it, but I jumped from the mast into the water.' And I said, 'Oh, I didn't know that. That's amazing that you did that. Were you okay?' Now I realize she was telling me all this because she wanted me to know that she wasn't scared when she died. But I look back at the conversation and I can kick myself."

*

"I feel a sudden rush of clairvoyance," Vic had written just days before her suicide.

> I am on the ledge. Heart pounding like crazy, my whole unbroken body shaking like a leaf. I have never felt fear worse than this. This is not dread, it's adrenaline. The finality of this moment makes me slowly breathe in and out. I have no last words, no note. Just a prayer. My first and last prayer to an invisible divinity: that I will die. I will forget that my bones will shatter, my heart will stop beating, my blood will rupture through a network of capillaries and arteries, my limbs will crack; bent in opposite directions; a repugnant, detached jigsaw puzzle. And I will fall like a book dropped from the top of an empty stairwell that lands with a deafening jolt on a concrete slab, penetrating the silence. Breathe in, breathe out. Close your eyes. Clench your fists. Fall. I open my eyes. I'm on my bed. Could have sworn I had done it.

*

Vic has a lesson for us, the living. "I guess I'm nothing more than another suicidal white girl," she writes. "Just another first-world brat succumbing to society's perfect illusions."

> I remember something J.K. Rowling wrote in the first Harry Potter book. That there were more important things than Hermione's affinity with books and cleverness. Like friendship. And camaraderie. Sadly, all that really matters, all that grownups are trying to drill into young minds, is success. If you are not successful, there is no point in existence. That's a pretty sad message to teach. But that is what's happening, whether we want it to or not. It doesn't matter if it's at the cost of one's wellbeing. Even if you are reduced to something barely functioning, but you pulled yourself up and are sitting there, telling your story in an Armani suit, that's all that matters. That you became a success story. You didn't wind up as road accident.

A few months before Vic died, over the Christmas holiday, she'd gone with her parents on their annual trip to New Zealand. "In the rush to catch the flight back to Singapore," Linda told me. "I'd left my good pair of glasses behind, which had sharp focus. It seems like a little thing, a dumb thing, but in the last few months of her life, I never saw Vic in proper focus." It feels like a metaphor.

We should all grieve for Victoria, for what might have been. Her mind, protected though she kept it from the people who loved her, was a brief, burning star whose brilliance has only reached us now that it has expired. The real tragedy isn't death, or loss, or even suicide, necessarily; those are the unwanted but inevitable companions of the human experience. The real tragedy, I think, lies in not being able to see a true star through the fog of our own passions and problems, even one burning there right in front of our eyes, hidden as it is in the blinding wilderness of the mundane.

"I knew she had a devastating wit," Linda told me about first reading her daughter's diary after her death, "but I had not realized the depth of her insight about the human condition."

6

to log off this mortal coil

All the inconveniences in the world are not considerable enough that a man should die to evade them; and, besides, there being so many, so sudden and unexpected changes in human things, it is hard rightly to judge when we are at the end of our hope . . .

Michel de Montaigne, *A Custom of the Isle of Cea* (1574)

When it first aired in March 2017, the controversial Netflix series *13 Reasons Why*—based on the young adult novel of the same name by Jay Asher—caused a huge stir in the suicide prevention community. And for good reason, I think. The gist of the story is that high school junior Hannah Baker kills herself because the people in her life have treated her poorly. While her locker turns into a colorful memorial decorated with Post-it notes about how much she'll be missed, we learn that before her death, Hannah recorded a series of audiotapes in which, as the title of the series alludes, she explains why she did it. Given that the target audience for the series is fifteen- to nineteen-year-olds, the age range especially at risk for copycat suicides, a lot of adults have been very concerned about this glamorized portrayal of a tragic young heroine who defeats her tormenters by taking her own life.

As it turns out, those "thirteen reasons" are in fact thirteen people who'd hurt or screwed Hannah over in various ways (a friend who blamed her for causing her breakup with her boyfriend, a stalker who leaked a private photo of Hannah kissing another girl and spread rumors of her being a lesbian, a boy who humiliated her while out on a date, another who sexually assaulted her, the counselor for not believing she was suicidal, and so on), and each of these people are to listen to the tapes in full. If any of them breaks the listening chain, then a friend of Hannah's, not on the tapes, has been instructed to make them public.

The network's decision to show the graphic scene of Hannah slitting her wrists in the bathtub also shrugged off all caution throughout the chillingly deliberate, and mostly silent, three-minute sequence. I gestated in the cultural womb of 1980s horror movies and can happily eat a steak dinner while watching someone get eviscerated by a chainsaw, and yet the hyperrealistic portrayal of Hannah's suicide, and her frantic parents' subsequent discovery of her lifeless body in the tub, left me deeply disturbed. Despite the vocal protests of mental health consultants, the TV executives ultimately ignored the science-backed warnings about the possible contagious effects of depicting suicide methods for adolescent viewers who might see such portrayals as step-by-step instructions.

In an opinion piece for *Vanity Fair* magazine written by Nic Sheff, who was one of the executive producers (along with the actress and singer Selena Gomez), we're presented with the traditional counterargument for how society should be handling this sensitive issue with teens. "There are many reasons why I'm proud to have worked on *13 Reasons Why*," wrote Sheff, a suicide-attempt survivor himself who says he was "quite surprised" at the negative reactions to the show by those in the field of suicide prevention. "But the thing I am most proud of, in all honesty, is the way we decided to depict Hannah's suicide."

> Facing these issues head-on—talking about them, being open about them—will always be our best defense against losing another life. . . . We need to keep talking, keep sharing, and keep showing the realities of what teens in our society are dealing with every day. To do anything else would not only be irresponsible, but dangerous.

Here's the thing, though. We live in an empirically informed world and, unfortunately, the data are in stark disagreement with Sheff's intuitions (and to be fair, many other people's intuitions as well) about what, exactly, is irresponsible and dangerous when it comes to this dicey topic of suicide and teens.

About thirty years before Hannah Baker was even a twinkle in Jay Asher's eyes, for instance, a similar show called *Death of a Student* appeared on German TV. In this six-episode series that aired during prime time on what was then one of the country's only two major networks, a despondent teenager throws himself in front of a train. In the first episode, we're shown the outcome of the suicidal act and the police investigation. The death scene is repeated at the start of every subsequent episode, each of which then goes on to unravel the sequences of events that led to the teen's suicide, from the perspective of his parents, fellow students, teachers, girlfriend and himself. Sound familiar? Just like the producers of *13 Reasons Why*, the laudable aim of the creators of this German series was to increase public awareness of teen suicide, to "get people talking."

It backfired. The result was a steep rise in teen suicide rates in Germany coinciding with the show's airdates, with young people using the same train-death method as the main actor. When the series was rebroadcast, it happened again.

*

Suicide contagion via the media is nothing new. In many cases, the connection is clear due to the copycat nature of the act. After Goethe first published his popular novel *The Sorrows of Young Werther* in 1774, a story in which the lovelorn protagonist—dressed meticulously in black riding boots, a yellow vest, and a blue coat—shoots himself in the head while at his writing desk, numerous young men dressed in similar attire, often with a copy of Goethe's book open before them or tucked away in their breast pockets, were found dead in the identical manner. There were so many casualties from this so-called Werther effect, in fact, that, for a time, the book was banned in several European cities.

In other cases, by contrast, suicide contagion may arise through far subtler media influences. One classic study examined the suicide rates in California between the years 1966 and 1973. The researchers discovered that whenever there'd been a front-page newspaper story about a suicide during that period, single-car road fatalities in the state increased significantly the following week. Because a surprising number of such incidents are thought to be suicides disguised as accidents—among other reasons, to evade life-insurance clauses—such findings seem to lend credence to the fear that suicide can operate like an infectious idea spreading among those most at risk. Any driver who has ever had suicidal impulses has felt that itching urge to swerve into non-being at some point in time, and presumably reading about suicide leads to "white bear" thinking (a classic study in social psychology showed that telling people *not* to think about a white bear only makes them think about white bears more).

Clearly, many people benefit from speaking openly about the terrible plague in society that is suicide. In fact, it's one of the main reasons I'm writing this very book. Having a certain degree of meta-cognition in this area—knowing how our suicidal minds are susceptible to being exploited by having knowledge of others' suicides—is an important line of defense against suicide contagion. If your guard's up when encountering suicide stories in the media, and you can analyze each case using the logic we've been exploring, then you're acting preemptively to protect your suicidal self against incidental and unprotected exposure.

It's admittedly complicated, though. The data paint a complex picture in which raising public awareness of suicide, and other well-meaning attempts to reduce the stigma attached to it, can make some already suicidal people more likely to take their own lives. Suicide is indeed among the leading causes of death in teens, for instance, but because death is so rare in this age group in general, conveying the message that teen suicide is common—when in fact the base rate of the behavior is quite low—efforts to "de-stigmatize" it may inadvertently lead some vulnerable adolescents to believe it's less unacceptable.

In July 2017, a brief study was published that may have confirmed, at least indirectly, critics' worst fears about *13 Reasons Why*. Through the application of "public health informatics," which is the practice of tracking and predicting health-related trends through the algorithmic use of publicly available social media data, a group of "infodemiologists" led by John Ayers revealed that Google searches for suicide-related terms and phrases rose dramatically following the series premiere. Compared to a representative time period just prior to the show's first airing, there was a nineteen-fold increase in all suicide search queries in the weeks afterward (March 31–April 18, 2017). Although some of that rise can be attributed to precisely the types of help-seeking behaviors the producers of the show wanted to encourage—phrases like "suicide hotline" (up 12 percent) and "suicide prevention" (up 23 percent)—the spike was also associated with far more worrisome Google searches. "How to commit suicide" shot up by 26 percent, "commit suicide" was up 18 percent, and "how to kill yourself" by a non-negligible 9 percent.

"In relative terms, it's hard to appreciate the magnitude of *13 Reasons Why*'s release," the study's coauthor Mark Dredze, a computer scientist, said in a news release. "In fact, there were between 900,000 and 1,500,000 more suicide-related searches than expected during the 19 days following the series' release." It's important to stress that these data don't reflect actual suicides. Still, similar informatics studies have found a clear correlation between suicide rates and online searching.

Despite our best intentions, then, even calling attention to the issue can be risky. In another study, a group of teens was shown a billboard with the simple message: "Prevent Suicide. Treat Depression. See Your Doctor." Most participants were unfazed by this well-intentioned appeal, but those kids who had a personal history of suicidality (thoughts or previous attempts) displayed maladaptive coping behaviors after being exposed to the sign.

Given that the Werther effect is especially insidious among young people, such results should give us pause when considering how to raise the issue with them. "The question of how and what to report in order to reduce the stigma surrounding suicide without promoting suicidal behavior," writes one research team, "while still providing information on risk and protective factors and coping strategies, including treatment resources, remains the foremost public health challenge regarding the media's role in suicide prevention."

It's not just a problem with impressionable adolescents, either. Or even traditional media. One man found dead of carbon-monoxide poisoning had an academic article about charcoal-burning suicide printed out beside his body—the same article I cited in chapter 3.

✷

Nearly two decades ago in the journal *Injury Prevention,* the sociologist Steven Stack, who has led the charge in advising the media on a set of professional guidelines for the responsible reporting of suicide, published a sort of meta-review of all data on the subject of suicide contagion that were available up to that point. The total sample comprised results from forty-two studies, each of which had investigated some aspect of the effects of media exposure to suicide stories on the rates of suicide in the general public. Although the findings weren't always clear-cut—with some studies indicating a significant copycat effect and others showing nothing at all—Stack was nevertheless able to discern a few reliable trends.

First, contagion is most likely to occur following media accounts of entertainment celebrities who've killed themselves (as opposed to other suicides in the public sector, such as politicians or similarly prominent figures). The largest known copycat effect occurred with Marilyn Monroe's alleged suicide in August 1962, where that month the suicide rate jumped 12 percent from its normal baseline.* Stack was also able to determine that a copycat effect is more likely to be found for stories that center on real, not fictional, suicides.

"People may identify with true-to-life suicides rather than make-believe suicides in movies or soap operas," argues Stack. "If Marilyn Monroe with all her fame and fortune cannot endure life, the

*Although not suicide contagion, a comparable statistic was found after the vehicular-accident death of Diana, Princess of Wales, on August 31, 1997. The suicide rate rose dramatically for both sexes the following month, but among females it shot up 34 percent.

suicidal person may say, 'Why should I?' " That doesn't mean that fictional suicides, such as that of Hannah Baker or Goethe's Werther, won't lead to copycats—clearly they can—but that, overall, they're somewhat less likely to have this effect. Additionally, by taking the perspective of the misunderstood decedent whose suicide leads to public outpourings of remorse, recognition, and appreciation, copycats may erroneously anticipate a future self who will enjoy a similarly positive social response after their own untimely deaths. (In the next chapter, we'll have a closer look at how our species' natural tendency to envision an afterlife can actually exacerbate the problem of suicide.) Not surprisingly, Stack also found that the extent of media coverage, which is measurable by the sheer number of media outlets reporting any given suicide, is positively correlated with copycats; front-page suicide stories or those that saturate the news cycle are obviously more of a problem than those given less attention by the press.

Perhaps the most revealing pattern uncovered by Stack, however, was that finding a copycat effect depends largely on how researchers define that term. If we include rates of *attempted* suicides rather than just *completed* suicides as evidence of suicide contagion, then we're far more likely to find that media exposure to suicide stories does, in fact, inspire copycats. News coverage of suicides may not always lead to an increase in actual suicide deaths, in other words, but they almost always lead to a notable spike in suicidal behaviors.

With decades of work showing that the Werther effect is real, media are strongly encouraged these days to adhere to a strict set of ethical principles when covering a suicide. Over at the suicide prevention website Reporting on Suicide (reportingonsuicide.org), there's a helpful list of do's and don'ts. Most would seem like commonsense, but as previous fiascos have amply demonstrated, that's not the case. *Do* "inform the audience without sensationalizing the suicide and minimize prominence (e.g., 'Kurt Cobain Dead at 27')." *Do not* "use big or sensationalistic headlines, or prominent placement (e.g., 'Kurt Cobain Used Shotgun to Commit Suicide')." *Do* "report on suicide as a public health issue." *Do not* "investigate and report on suicide similar to reporting a crime." *Do* "use school/work or family photo." *Do not* "include photos/videos of the location or method of death." Those sorts of things.

Here in New Zealand, where attempted suicide was, embarrassingly, considered a crime until 1961, it's unlikely you'll come across an unambiguous case of suicide in the press today, despite the country's high suicide rate. The media talks about it in the abstract, as when a desperate family pleads for information about their depressed loved one who was last seen heading toward some remote, windswept beach that's flanked by 110-foot cliffs, and you can piece it together. But there's no direct talk of a suicide when the body is eventually found and "no foul play is suspected." Given Stack's findings, that's probably a good thing.

Back in the mid-nineteenth and early twentieth centuries, however, such detailed stories were anything but difficult to find in this country. From 1865 to 1901, the *Evening Post* newspaper, based in the capital city of Wellington, published 8,601 suicide stories—averaging more than one a day—and typically with sensational, attention-demanding headlines. A quick sampling from that era brings up titles such as "Extraordinary Suicide of a Girl," "Strange Double Suicide," and "Dramatic Suicide—Dive into Molten Metal." See what that kind of wording does to your cognition? Now you're curious, which is just human nature; yet this curiosity is also a potentially deadly lure. (To ease your cognitive needs, the girl tied a rope to a flax bush and jumped off a bridge so that she both drowned and hanged, the double suicide was a pair of young widows who shot themselves on board the liner *Luciana,* and the molten-metal dive was by an underpaid steel worker who jumped into the glowing red mass while smoking his cigar.)

Down in Dunedin, where I live, it was the same. I think of a man named John Mair almost every day. Why? Because on my drive to work, I go right past the very spot where he shot himself in the head at eight o'clock in the morning on November 9, 1920. I can tell you quite a lot about the scene ("He was lying face downward on the pavement . . . on the footpath fully dressed, bleeding from a wound in the right temple, and there was a revolver lying about two yards away. It was fully loaded with four ball cartridges and one had been discharged. . . . There was a wound about the size of a shilling on his right temple and a slight abrasion on his left cheek, probably due to a fall. . . .") and about the circumstances leading up to it ("Deceased had been keeping well, but depression had come on him since his father's death nearly two years ago. . . . [L]ately he'd become subject to delusions that the house he was living in was going to be sold. He suffered badly from insomnia. His relatives had not realized to what a pass things had come with him. . . ."), thanks to the colorful historical archives of the local newspaper and my penchant for useless detective work.

✷

There has, of course, been one major development in the years since Stack published his important meta-review of the effects of television and print media on public suicide rates—and as you've probably surmised, it starts with an "i."

Just as the internet has changed every other aspect of human social behavior, it's altered the suicide landscape dramatically. It's given everyday people, including those who are actively suicidal and therefore more easily influenced, unprecedented access to material that, not so long ago, would have been off-limits to all but psychiatrists, police, and coroners. Typing the word "suicide" into the Google search bar today yields no less than 288 million hits, and although the most common content online is about suicide prevention, it's certainly easy enough to find—or stumble upon—less benevolent material. One content analysis focusing on a random 373 websites devoted to the subject of suicide revealed that 31 percent were best characterized as suicide neutral (neither advocating nor condemning the act), 29 percent were anti-suicide (offering links to mental health resources and encouraging visitors to seek help), and 11 percent pro-suicide (promoting suicide as a solution to problems and often including detailed information on various methods).

For some online users, simple curiosity may be behind visits to websites with disturbing images or those featuring quasi-philosophical, rather than preventative, discussions about the moribund subject. A 2013 study in the *Journal of Medical Internet Research*, for instance, tells us that search terms used to access such pro-suicide sites included "committing suicide with a gas oven," "pictures of murder by strangulation," and "photo of a severe burn." Still, even if we grant that a fair proportion of these ill-advised sojourns on the internet are motivated by puerile interests,* the worry—and I think it's a legitimate one—is that there are many people genuinely on the edge who will find themselves poring over this lugubrious material as well. A recent cross-cultural study by a team of Finnish researchers found that some of the most likely visitors to these pro-suicide websites were in fact recent victims of online bullying.

Many of these sites, of course, prefer to trade under the heading "pro-choice" rather than pro-suicide. At one controversial Swedish website, which has been around since 2005, visitors are told that: "The purpose of this page is to try to report all suicide methods existing in this world as objectively as possible. The idea is not to encourage anyone to end his life, but to spread information about the different methods." Despite the lip service, the content makes it clear that the site operators disagree with Camus's famous sentiment ". . . in the end, one needs more courage to live than to kill himself." Their original dedication was to ". . . everyone who has had the courage to flee from this hell by killing themselves." We're then given a helpful to-do list when mentally preparing ourselves for suicide—you know, for getting over that pesky will-to-live thing. Such dangerously irresponsible advice, which includes the suggestion to repeatedly visualize one's death, as Vic McLeod so tragically did, gas-pedals that most precarious disinhibition stage. Also, for those so inclined, there's a generous menu of options for how to kill yourself, neatly categorized under labels such as "Toxins," "Sharp Blades," and "Animals." Here's a rather exotic one that used to be listed in the "Jumping from Heights" section, for instance:

*I couldn't bring myself to watch such upsetting footage now, but I can't tell you how many times I watched the mondo horror film *Faces of Death* as an overly curious twelve-year-old in pre-internet America.

> This method is a variation of jumping off a bridge, but more on the artistic side. You need a high bridge, rope, and piano wire. Cut the rope into various lengths. Each length must not exceed the height of the bridge. Tie one end of the ropes and piano wires to the bridge. Tie the other ends of the ropes to different parts of your body, such as thighs, calves, torso, etc. (Do not forget your genitals.) After that, tie the piano wires around your joints. When you then jump from the bridge, different parts of your body will be cut off because of the loops of piano wire, and your body parts will be left dangling in mid-air, suspended by the ropes. If you do it right, you'll end up with just your torso hanging from the neck over the sea, highway or ground. If you're lucky, someone will photograph you, and you'll be immortalized as a work of art.

This is a poor assessment of the market, in my opinion. Suicide will always flop as performance art. The critics will write you off as a desperate bore with no future, and the audience will despise you for your conceited indifference, if not outright antipathy, to their life-affirming worldview. You'll be forgotten in days.

Oh, and also, *ouch.*[*]

✷

If there was ever any question about the role of the internet in suicide-related trends, it was finally put to rest in the nondescript

[*]In a thoughtful analysis of the motivating psychology behind such websites, Michael Westerlund, a professor of media and film studies, suggests that these dark web forums can, oddly enough, provide a salubrious sense of purpose to people trying to make sense of an absurd existence. "The expressions of violence and morbidity can be seen as a total rejection of cultural and societal demands for the discipline and sublimation of instincts," he writes, ". . . opposing everything the dominant culture finds sacred. . . . Although construction of the pro-suicide approach can in many ways be seen as a destructive activity, it does, at the same time, constitute a meaningful and sense-making activity for its protagonists. . . . So, paradoxically, this production and preoccupation with violence, death and suicide ultimately concerns the aspiration to create meaning and purpose in life." Yet the same, of course, can be said for any antisocial endeavor. And if your primary source of meaning is derived from suggesting that other people kill themselves, you're basically a purpose vampire, stealing their ability to find meaning in life to nourish your own.

Japanese industrial town of Iruma on February 11, 2003. In chapter 3, we saw how the strange death of Jessica Choi yuk-Chun—the young insurance executive from Hong Kong who killed herself using what was then an unheard-of method of burning charcoal in a sealed-off room—led to a copycat suicide epidemic in several Asian countries. In the Saitama Prefecture of Japan five years later, three young people, complete strangers to one another, would meet on a pro-suicide website and enter into an online suicide pact to die by charcoal burning while holding hands. The Japanese media was quick to see this as a shocking new form of suicide, *net-jisatsu* (internet suicide) and, in an act of dumb irony, proceeded to then cover the incident extensively. In the months following the detailed reporting of the case, there was, as you'd expect, a proliferation of group charcoal-burning suicides of the same *net-jisatsu* variety.

More recently, suicide by hydrogen-sulfide gassing has replaced charcoal-burning as the nation's leading method. Most experts trace this shift to a 2008 suicide pact involving three other young Japanese people who'd met online. Once the media reported on that trio's decision to die together by filling the room with hydrogen sulfide gas, an obscure method at the time, more than a thousand such hydrogen-sulfide suicides would accumulate before the end of the year (thirty-five times the number reported for the previous year). Google searches for the term "hydrogen sulfide" shot up fifty times in the initial weeks after the suicide pact was reported and surpassed the search volume even for the word "suicide." The online publication of a handy set of instructions for how to make the deadly gas at home didn't help matters.

I guess this is one of the few situations—rigging a makeshift suicide apparatus or preparing lethal chemical concoctions—in which being completely incompetent at anything requiring a sliver of technical know-how or manual ingenuity can be, ironically, a great lifesaver. Personally, I'm terrible at such things; I can't even put together a chair from Ikea. Suicide by gas? Highly unlikely, in my case.

In fact, as a way of offering you a bit of a breather from such heavy material, let me give you an example of just how bad I am with death-related contraptions. Thinking I'd probably lose a finger setting a mousetrap, I once made the unforgivable mistake of using a glue trap to catch some thieving vermin in my pantry. Pelleted remainders of the evacuated bowels of rodents mixed with the nibbled cardboard shavings of what was once an unmolested rosy-cheeked portrait of the smiling Quaker Oats man littered my shelves. Disgust is a powerful emotion, especially for suburbanized germophobes desperate to convince themselves that they're *totally* comfortable living all alone in the country. It's so strong that all signs of sensibility, and above all compassion, evaporate when it rolls into one's mind. Anyway, didn't I hear stories about mousetraps sometimes acting more like a guillotine? The thought of a severed rat head next to the peanut-butter jar—besides oatmeal, my other food staple in those days—was just too much. "Looks harmless enough," I foolishly thought, suckered in by the image on the glue trap box of a villainous cartoon rat tugging at his leg. "Probably won't catch anything anyway."

To my lasting regret, I was wrong. What the manufacturers of these inhumane products don't tell you is that once you make eye contact with the pitiable creature pinned to the tarry surface of the board, you've shaken hands with Satan. I'm not sure what I was expecting to see when I opened the pantry door the next morning, but it wasn't the cute little brown-eyed field mouse I found. I stood there for a while feeling as shameful as a criminal, looking down at this poor animal that was flattened out as if surgically prepared for a vivisection, its furry little ribcage heaving, its whiskers twitching, and I apologized profusely. I apologized for the hopeless situation I'd put it in and, even more, for what I was about to do to it.

The idea of simply disposing of it in the trash, alive, was a nonstarter. Allowing it to starve to death on the side of a dirt road, boiling in glue in the apex heat of the Arkansas summer, seemed unspeakably cruel to me. Sure, I'd squashed plenty of insects in my day without the slightest nip of a guilty conscience. Once I even flushed a live spider down the toilet. But this felt—and I don't know if it was what I imagined was a pleading look in the animal's eyes or my Disney indoctrination as a child—different. "What the hell, sir?" it seemed to be saying. "*Jesus*. I just wanted a bit of your precious oatmeal."

I had to put it out of its misery. And fast. But how? I didn't even own a pellet gun.* True country folks would probably employ a hammer or just stomp it with the heel of a boot. As any forensic investigator will tell you, though, such proximate methods of murder are indicative of hatred, or at least passion. Of that I had none. If I ever wanted to sleep again at night, I'd need distance. Yes, this minor beast's ilk may have been partly responsible for pestilential deaths throughout history, but not the innocent before me.

Googling, I learned that to prepare mice for snake food, herpetologists use a purpose-built CO_2 chamber. "A fill rate of about 10% to 30% of the chamber volume per minute with carbon dioxide," the National Institutes of Health inform the budding rodent executioner, "added to the existing air in the chamber should be appropriate to achieve a balanced gas mixture to fulfill the objective of rapid unconsciousness with minimal distress to the animal." Considering that I was too thumbsy to rig a 75-cent spring-loaded mousetrap, the idea of constructing such an apparatus within the next millennium, let alone the next five minutes, never really took flight. Gas chamber? I might as well build a mouse-sized electric chair. Without belaboring the unpleasant end to this poor creature's conundrum, suffice it to say that it was a merciful drowning. And after emptying the bucket with its sad contents into the weeds behind the cabin, I turned to the only witness to my crime, my faithful dog, Kit, who was getting on in years at the time. "I think we need a cat," I told her.

(You remember my "suicidal" cat Tommy, right? Well, that's his origin story.)

*It was, I knew immediately, a lost cause. I contemplated trying to peel it free somehow but cringed at the injuries likely to result from such a delicate operation. These days, if you Google "how to free a mouse from a glue trap," you'll get several dozen optimistic how-to websites, even a helpful YouTube video of the process, instructing you to, say, dab the affected body parts gingerly with cooking oil before gently massaging the frenzied animal's glued bits free (while also managing to avoid getting bitten and contracting a disease). I've no doubt this works in some cases. But even if such internet wisdom had been available then, this patient's condition was grim. Not only were all of its tender appendages fixed implacably to the surface of the board, its entire underside, hair and skin, was also stuck.

I took the scenic view to my point, but regarding some forms of suicide by gassing then—anything involving chemical equations or tubes, chambers, cylinders, gauges, and other mechanical bits and pieces that my frontal lobe won't or just can't handle—a level of engineering skill, or at least a capacity to follow technical instructions, is often essential. So, if you're anything like me, don't fret over your mechanical ineptitude when it comes to suicide: it may be a godsend, especially with so many methods being so easily researchable online. We might die of old age after all.

The trouble is, with incidents like those *net-jisatsu* hydrogen-sulfide suicides in Japan, it's clearly also all too easy to find a more capable person to rig your death . . . and to die with you. The internet is both a vast information repository and a social watering hole for the suicidal. This is particularly perilous for those who are, in the words of one survivor, "too lonely to die alone." The young woman who'd used this expression to describe herself had been frequenting pro-suicide sites, looking for what the Japanese sociologist Hidenori Tomita calls an "intimate stranger," an online presence whose anonymity becomes the basis for intimacy. "Persons who wish to commit suicide may not want friends to die with them," explains the anthropologist Chikako Ozawa-de Silva, "due to their friendly care or emotional attachment; these same persons may not want to die with an enemy, due to feelings of aversion. A stranger, however, causes neither problem, while still creating the safety of a group environment and the comfort of taking a communal rather than an individual choice."

Such cybersuicide pacts appear to be especially common in collectivist Asian cultures, where dying alone is heavily frowned upon. In South Korea, which has one of the highest suicide rates in the world (24.1 out of 100,000 in 2015), such deaths now account for almost one-third of all suicides. But they're certainly not unheard of elsewhere. England's first publicized case of a cybersuicide pact was in 2005, when the bodies of a twenty-five-year-old man and a forty-two-year-old woman, who'd apparently never met in person before their deaths, were found slumped over in a vehicle in a South

London parking lot, with a tray of—you guessed it—smoldering charcoal between them.*

✷

Advances in social media technology mean that livestreaming one's own death (broadcasting it over the internet in real time, typically after clearly communicating one's intent to do so for a watchful online audience) has introduced an unprecedented problem for suicide prevention. Previously, such shocking, and mercifully very rare, events would happen only on live television. The most notorious case happened in 1974, when a mentally disturbed young news anchor named Christine Chubbuck announced to her Sarasota, Florida, viewership that "in keeping with Channel 40's policy of bringing you the latest in blood and guts, and in living color, you are going to see another first—attempted suicide" before lifting a revolver from beneath her desk and fatally shooting herself behind her right ear.

Today, any suicidal person with a webcam and an exhibitionist streak can do the same, and unlike Chubbuck's death (the footage of which was never aired on television again), these upsetting death scenes are virtually impossible to erase from the public domain once they're caught in the web.† There have always been those in search

*Prior to the internet, the traditional profile of a double suicide was very different. Typically, they involved elderly codependent husbands and wives facing unemployment or serious physical illnesses. And historically, they were quite rare. Between the years 1955 and 1958, there were only 58 such cases out of a total of 20,788 suicides in England.

†Although his intent remains somewhat unclear, the first livestreamed suicide, or at least suicidal act, was that of Brandon Vedas, a twenty-one-year-old from Arizona who logged onto an IRC (Internet Relay Chat) through the website Shroomery.org in the early morning hours of January 12, 2003, and announced to the other chat members that he'd set up his webcam for them to watch his "grip of drugs." After ingesting several psychedelic mushrooms, Vedas downed an entire bottle of methadone, two Vicodin tablets, a handful of the tranquilizer Klonopin, and innumerable anti-anxiety and anti-insomnia drugs. "I told u I was hardcore" was the last thing he typed before seizing and lapsing into unconsciousness. His mother found his dead body the following day. Reactions from the other chat members was mixed. Some egged Vedas on, others urged him to stop, and several debated calling 911 or trying to track down his IP address. In the end, though, all anyone did was watch him die.

of fame who could find no other route but through the infamy of staging a grandiose death; the internet now offers these desperate people the amber to freeze their darkest moment in time.

Although online suicides are still uncommon, they appear to be growing in frequency, particularly with easy-to-use video-streaming platforms such as Facebook Live, which was unveiled to the public in January 2016. In April the following year, a twenty-three-year-old college student live-streamed the final moments before his fatal leap from a nineteenth-story window of a suburban West Mumbai hotel. Just a few weeks later, a thirty-two-year-old lawyer in nearby Sewri tried to do the same from the eighteenth floor of a building that was under construction, but got intercepted by police after they'd been alerted not by anyone from her online audience, but by a concerned nearby resident who happened to notice the woman acting strangely.

Others haven't been so fortunate. And these live-streamed deaths are stacking up. Here's a small sample from 2017 alone: forty-nine-year-old James Jeffrey from Alabama shot himself in the head after a nasty breakup; Jared McLemore, a thirty-three-year-old Memphis musician, doused himself with gasoline and set himself on fire in front of the bar where his former girlfriend, a soundboard artist, was working a gig; fourteen-year-old Nakia Venant from Miami fashioned a noose from a scarf and hanged herself while her verbally abusive mother, watching the live footage, typed that she was just a "crying wolf . . . seeking attention"; Erdogan Ceren, a twenty-two-year-old Turkish man upset about his girlfriend dumping him, told his viewers, "No one believed when I said I will kill myself—so watch this," before pushing the barrel of a shotgun against his stomach and shooting himself; and on the personal broadcasting site Live.me, twelve-year-old Katelyn Davis of Georgia, an active blogger with a history of self-harming, kept the camera running as she hanged herself from a tree in her backyard, alleging abuse by a male family member as the reason for her suicide.

Why would anyone want to kill themselves in front of a live audience? "A suicide is not just about dying," writes Michael Westerlund, an expert on suicide and the internet. "A suicide also communicates something to the world around the person, such as that the person feels unloved and outcast."

> The suicidal person achieves something with the act, something that the recipients cannot escape. It is a way to regulate the social environment. The suicidal act can be seen as a powerless person's weapon to influence the outside world in a way in which the recipients are deprived of the ability to speak back, which is a fundamental point.

In 2015 Westerlund and his colleagues Gergö Hadlaczky and Danuta Wasserman published a unique case study in the *British Journal of Psychiatry*. In this piece, the authors analyzed the complex online social dynamics surrounding the death of twenty-one-year-old Marcus Jannes. On October 11, 2010, Jannes, a college student living in Stockholm, started a thread titled "Hanging" on the Swedish internet forum Flashback.org, which prides itself on being a free-speech venue for its one million (mostly male) members to talk openly about subversive stuff—extreme political views, hacking, illegal downloading, drugs . . . those kinds of topics. "I have now decided to kill myself by hanging," wrote Jannes in his initial post:

> I have softly tried to strangle myself and saw how that feels. Took some painkillers a few minutes ago, now waiting for it to start working. Have turned on my webcam with a program that makes a screenshot every 2 seconds and put up an FTP [file transfer protocol, a standard way of transferring computer files on the internet] where the images will be available . . . will post the IP, port and login details before I do it.

The first two replies to Jannes's straightforward post, which left little room for misunderstanding, show just how varied were the viewers' reactions to such a startling announcement.

"Good luck then!" wrote the first person.

"It can't be that bad," chimed in the second. "When everything is at its worst, it can only get better."

Over the next hour or so, Jannes went back and forth with a

small group of others who subsequently joined the fray, discussing with them his imminent suicide. Some thought he was just trolling them—trying to get a rise out of the hard-to-shock Flashback community. But he gave little reason for doubt. When asked why he wanted to kill himself, Jannes wrote: "I have Asperger's syndrome/high-functioning autism. Am overly vulnerable (emotionally) . . . have rather poor social skills, which makes me a somewhat lonely person."

Recent research has confirmed the previously anecdotal evidence that those on the autism spectrum, like Jannes, are at a significantly increased risk of suicide, presumably due to the social difficulties that so often go along with the cognitive disorder, such as exclusion, unemployment or workplace difficulties, and trouble forming close relationships and developing romantic bonds. A 2014 study from the *Lancet* found that of 374 British men and women with Asperger's syndrome, 127 of them—a whopping 35 percent—had active plans or had made previous attempts to kill themselves. Compared to the general population, they were *nine times* more likely to self-report suicidal ideation.*

"It has always been a scary thing to kill oneself, as you might understand," wrote Jannes,

> but after I tested strangling myself with my hands, so that the blood vessels in the face began to break, it did not seem so scary anymore, but more filled with peace, like I finally would come to rest.

Looking at the exchanges during this critical period when intervention was still possible, twenty-one messages were directed at

*Interestingly, many of these patients were suicidal even though they claimed not to be depressed, which the investigators speculate may be due to the difficulties faced by many people with Asperger's in describing or verbalizing their subjective emotional experiences (*alexithymia* is the technical term for this). In fact, Jannes seems to have fallen into this not-depressed-but-suicidal category. Either that, reason the investigators, or these puzzling findings "could suggest a different process for suicidal ideation in Asperger's syndrome than for other clinical groups."

Jannes. Of these, nearly half could be characterized as encouraging him to do it.

"You stupid fuck, strangulation is no pleasure," wrote one such person. "Don't you have a car? Carbon monoxide rules."

Another jackass weighed in with "you won't dare, you are too cowardly."

And then another: "You're just a faker, go and hang yourself."

More positively, and hopefully allowing you to regain at least a smidgen of your faith in humanity, seven posts were attempts to discourage Jannes from killing himself.

"Don't do it, there are other solutions," wrote a far kinder person than those others.

"Can't you tell us a little about your life?" pleaded another.

The investigators noted that these supportive gestures were beginning to get to Jannes. "Starting to feel that I'm about to change my mind about killing myself," he writes. "So I have to hurry up a bit." Alas, he was committed to dying, and it's hard to know what role, if any, those inciting him played in the event.*

In his final post, which was made just after the video showed him

*In a rather prescient article from 1959, the psychiatrist Joost Meerloo referred to such incidents as *psychic homicides,* and in doing so it's almost as if he foresees the coming of the internet—or something like it. "The act of suicide," wrote Meerloo in this nearly sixty-year-old piece, "may be the follow-up of a command and verdict of a proxy, a person the victim identifies with. . . . The act is legally not punishable yet [but] in the age of encroaching technology and growing community pressure, resulting in weakening ego, decreased self-esteem, and diminishing personal responsibility, these attacks on a person's will and integrity become more and more relevant." Those commenters who'd encouraged Jannes to take his life were briefly investigated by Swedish police, but no charges were ever filed in the case. However, seven years later, the successful prosecution of Michelle Carter in June 2017 in Massachusetts' high-profile "texting suicide case"—in which the then-seventeen-year-old defendant was found guilty of manslaughter for sending her suicidal eighteen-year-old friend Conrad Roy III a barrage of demanding texts urging him to kill himself—set an important precedent for these types of crimes. In that case, Roy, a newly minted marine salvage captain with a history of anxiety and depression, pulled into a quiet part of his local Kmart, hooked up a portable generator he'd acquired for the purpose, and filled the cab of his pickup truck with carbon monoxide. As he began to asphyxiate from the toxic fumes, he panicked and left the vehicle; Carter texted him to get back inside and finish the job. He did as she told him.

in sweatpants and a T-shirt standing in his apartment and carefully draping some computer network cables from the arch of a doorway, he writes: "all right, let's do it."

And tragically, he does.*

✷

With access to intimate strangers willing to die with us as suicide partners, a worldwide voyeuristic audience perversely ready to watch us do it, and detailed information about how to complete the act "successfully" at our fingertips, the internet has clearly troubled the already troubled waters of suicide prevention. Yet another axiom of the electronic age compounds the problem still, which is that the dead—especially the youthful dead—will so often be memorialized online. These virtual breviaries of adoration can be dangerously seductive to young people who covet social media approval.

Lindsay Robertson, a New Zealand–based researcher in preventive and social medicine, analyzed the content of online memorials

*Westerlund and his coauthors' analysis didn't end there, however. The researchers also scrutinized the group's response *during* and *after* Jannes's suicide. (The thread was left open for several hours after he was dead, during which time thousands of additional posts were made.) After debating whether or not they should contact the police for fear of implicating themselves, some of the viewing forum participants eventually tracked Jannes's IP address and alerted local authorities. By then, however, he was long dead; in fact, the video was still streaming when first responders arrived an hour after he'd hanged himself. Despite the graphic images, many posters still believed that it was a fake. And although many saw the incident as a tragic and terrible incident ("Rest in peace. I was too late to write something that would make you change your mind. Very sad way to end one's life"), a significant minority wrote that the suicide was exciting, interesting, or funny ("Call me sick, but have never laughed so much in my life"). Others now joining in attributed blame to those who'd incited Jannes ("I think that you should be ashamed, you who wrote 'good luck,' 'you will never dare,' etc., the fact that you can write something like that to a guy who obviously does not feel well is completely hellish, it could be those words that gave him the motivation to pursue it. Frankly, you have been involved in this"), and a considerable number—including the forum moderator—reasoned that it wasn't anyone's place to interfere with his suicidal intentions if that's what he wanted ("Why stop the guy? If he doesn't feel like living any longer, it is up to him to make the decision whether to do it or not").

to young suicide victims. In a 2012 article for the journal *Crisis*, Robertson and her colleagues wrote:

> These pages typically included photos or slideshows of the deceased young person, videos, poems, statements about suicide, and messages posted by friends and family members acknowledging the person's life, all of which could be viewed by everyone who had access to the site. The content of material posted to the sites was overwhelmingly positive toward the person who had suicided, celebrating their life, describing the positive qualities of that person and how much they would be missed.

One of the victims in this memorialization study also had an elaborate funeral that was filled to capacity owing to the details of the event being circulated widely across the decedent's social media accounts. "Sources reported that young people were heard commenting on how impressive the funeral was," Robertson and colleagues wrote, "and how they hoped their own would be similar." This envying of the dead is the reason behind suicide-reporting guidelines that caution the media against featuring photos or videos of grieving family and friends at memorials and funerals.

This isn't to say that we shouldn't mourn, or indeed celebrate, those lost to suicide, only that it's a dangerous mix when the digital age puts such a premium on clicks, views, likes, followers, and comments, and the only perceived guarantee of receiving such attention (or perhaps just halting an assailment of negative judgment) is death. After all, it's hard to find a kid—or an adult, for that matter—without a phone attached like a barnacle to their hand. And that's a problem when paired with a few other fast facts. First, among older adolescents, the use of social networking sites such as Twitter, Facebook, Instagram, and Snapchat has been found to correlate positively with mental health issues, including depression, anxiety, and suicidal thoughts. Second, these social-networking sites have been labeled "addiction prone technologies." And finally, cyberbullying—which is typically defined as "the use of email, cell phones, text messages, and internet sites to threaten, harass, embarrass, or socially exclude"—affects an estimated 20 percent of contemporary middle and high school students.

From a mental health perspective, experts believe that this form of victimizing can be even worse than its "real world" equivalent. Cyberbullying can occur day or night, being brought into the otherwise safe haven of the home and anywhere else where there's a cell phone or internet connection. Nasty messages, embarrassing pics, and public smears are instantly distributed to one's peers and often across a wide network of complete strangers, leaving victims not only worried about their reputations but also fearing for their safety.

One recent meta-analysis included an enormous sample of nearly 285,000 participants aged nine to twenty-one years and found that being the victim of cyberbullying was more likely to lead to suicidal thinking than getting stuck in the crosshairs of old-school traditional bullies. (They often go hand in hand, needless to say, with bullied children being victimized both online and offline, but these data pan out even when you isolate the instances to "pure" cases.) Cyberbullying has become so problematic that the Centers for Disease Control and Prevention has designated it "a serious public health threat" and put out warnings to parents.

I think I breathe a collective sigh of relief on behalf of an entire generation of former teen misfits who traveled through those unsteady pimply years before social media came along and upended absolutely everything. I don't know if I'd have survived my gay love-letter affair from before if email had been around then, and I also recall an incident in which I wrote a long and steamy note (are you beginning to see a pattern to my problems?) to my "beard" when I was a bantamweight high school freshman. That note got passed around, Xeroxed, and, by the end of second period, was pin-tacked in multitudes to the bulletin boards in the cafeteria and hallways. My memory is good but not that good, so it would be misleading to try and reproduce the entirety of it for you now. Still, I recall the gist of it, and one thing I'm certain of is that it included some especially vivid lines about how this girl "looked like a model" and that it would be both an honor and, indeed, a pleasure if she allowed me to floss with her pubic hair. "Was that so much to ask?" I pleaded. As I so often did in those days, I went a bit overboard in my faux heterosexual enthusiasms.* Creepy and weird, yes, but the point is, humiliation tactics were much shorter-lived when I was a teenager. If I'd have graduated a decade later, I'm sure you'd be able to find a copy of that embarrassing note in the bowels of the internet and read it verbatim.

Regardless of our age, it's vital to realize that the online world exploits our evolved social psychology in predictable ways. Philippe Rochat calls rejection "the mother of all fears," and for an animal whose emotions are so tightly bound to what others think of it, the internet can be a true incendiary device for igniting our deepest anxieties. One of the best examples illustrating our extraordinary sensitivity to being "e-ostracized" comes from the social psychologist Kip Williams's cyberball experiments. In these studies, you're one of three people playing what seems to be an innocent ball-toss game. It's easy to picture the basic setup. If you're looking at the computer screen, you'd see your outstretched virtual arms and two figures—one to your left, one to your right—with the three of you tossing a ball. In the standard design, you're led to believe that these other two people are participants in the study just like you, playing the game from some other location. In fact, they're not real at all, and you're the only genuine subject in the experiment. The first few minutes go by nicely; the rules of social etiquette dictate that each of you tosses the ball to the next person. Suddenly, these other two characters, following a preprogrammed script, begin ignoring

*A biology teacher eventually came upon the ridiculous and coarse note, tore down all of the copies from the bulletin boards, and saved one to share with the vice principal. The vice principal, in turn, placed a phone call to my mother, who was busy stuffing envelopes at a medical billing company up the road, and regaled her with her son's morally questionable thoughts by reading aloud the offending matter. But—and this is just hearsay, and she is long dead—from my mother's perspective, the vice principal's reading was not without some discernible pleasure and may have even been laced with, what, vague compliments about my writing? And so my parents, the rare breed who'd prefer a literate sexual deviant over an idiot for a son (not to imply that the two are mutually exclusive, but my father, a peripatetic glue salesman who'd majored in English and who had a penchant for bisexual poets, was uncommonly nonjudgmental), praised me for the well-written note, then let me off with a halfhearted admonishment to never, ever again speak so impudently about a person's pubic hair . . . a rule I've happily violated many times in the years since.

you . . . in an apparent snub, they're now throwing the ball only to each other.

A few occurrences of this are enough to generate negative emotions such as sadness and anger in participants, and that's true not just for "overly sensitive" types, but pretty much everyone. In fact, our detection of social rejection is so highly attuned that the cyberball effect holds even when participants are told in advance that it's just the computer doing the ostracizing. It also works when you think the other two people are from a despised outgroup, like neo-Nazis or KKK members. And it works when you're told there's a financial incentive to being excluded and being included in the game incurs a penalty fee. One of the more interesting studies in this area showed that being a cyberball reject impairs health-related willpower; after the game, participants are more likely to binge on cookies and to refuse to drink an unpleasant-tasting beverage they've been told is good for them.

Being ignored online is obviously a disagreeable experience in its own right. When you think of cyberbullying victims who harm themselves, or even take their own lives, the experience is probably more akin to having the cyberball thrown right at your poor cyberface and breaking your damn cybernose. Over and over and over again. "How can an otherwise perfectly socially integrated [person] commit suicide after being the target of attacks on a virtual social network?" asked Matthieu Guitton, editor of the journal *Computers in Human Behavior*. "The 'social network' is not the platform," he writes, "but the community of people interacting between themselves through this platform. Some people may use this platform while forgetting—or sadly, not realizing—that behind each account is sitting one of us, another human being with his strengths and weaknesses."

*

The internet is a manifestation of human nature, and because of its unique capacity to bridge formidable social divides, it's important to emphasize that it summons not only the worst in us, as we've seen, but also an astonishing amount of good. The positive influence of media (social or otherwise) on suicide prevention was dubbed the "Papageno effect" by the Austrian suicidologist Thomas Niederkrotenthaler in a 2010 article in the *British Journal of Psychiatry*. The literary, and literal, opposite of the Werther effect, it is so named after a scene from Mozart's opera *The Magic Flute*, in which the lovesick character of Papageno is persuaded by the spirits of three young boys not to take his life when he becomes desperately suicidal.

Although it's more difficult to measure the positive influences of social media on suicide prevention—how many lives have actually been *saved* by having relatively effortless communicative access to concerned others isn't easy to ascertain—all of the major social media companies now have fairly simple mechanisms in place for users to anonymously report posts that threaten or indicate self-harm or suicide. The flagged posts are triaged by a dedicated team of trained staff who work to quickly assess the relative risk for each submitted case. Once the online alarm button is pushed by a member of the public, network responses range from no follow-up needed, to reaching out to the original poster and helping them liaise with a trained crisis counselor or someone from their nearest suicide prevention organization, to—in the most urgent scenarios—sending first responders to the individual's home.*

Internet relationships are often, for better or worse, the most significant, and sometimes the *only*, meaningful social ties that a person has. For those who lead busy lives and have few opportunities to regularly share their problems with like-minded others dealing with similar daily stressors, for instance, online support systems can be lifesavers. Take veterinarians, who have one of the highest suicide

*Some companies, including Facebook, are currently experimenting with artificial intelligence technology to automatically detect suicidal users (by identifying words and phrases in posts such as "I just wish I was dead" or "I want to kill myself"), but so far these efforts have been hit-and-miss. "Using AI to identify people who are thinking about suicide, and then reaching out to them, may be very helpful," Michael Graziano, a neuroscientist, told the *Deccan Chronicle* in 2017. "But even if it helps to some degree, for some people, it obviously won't solve the whole problem, so you'll always be able to point to some spectacular tragedies. . . . I don't think giving emotions to AI would make any obvious difference to that effort, at least not right now. Human beings are good at emotions, and yet not very good at suicide prevention."

rates of any profession. Recently, I spoke with Carrie Jurney, a veterinary neurologist based in San Francisco and one of the moderators of the private Facebook group NOMV (Not One More Vet), which claims an impressive 10 percent of all U.S. veterinarians in its membership. The formation of the group was inspired by the 2014 suicide of Sophia Yin, a popular vet and animal behaviorist who'd been in the public eye for years as a television host on the Animal Planet channel, as well as a writer, speaker, consultant, and peer educator. "She was a very positive member of the [veterinary] community," Carrie told me. "She was instrumental in what we call the fear-free practice movement, which is about using positive reinforcement to train your pets and making your practice a happier place. She worked a lot. She focused on interactions with children and dogs. Stuff that is very feel-good. And so, I think when she died by suicide, we all just took a breath. . . . Because she was such a positive force, it was kind of a gut kick, you know, like if it can happen to Sophia, it can happen to any of us."

The fledgling community quickly grew through word of mouth, meeting what was, apparently, an enormous unspoken need for people in this field. Carrie told me that there are only two rules. Number one, you must be a veterinarian. And two, you cannot be an asshole. "I know that second one is a little vague, but we've expanded it as the group has gotten bigger and the dynamics change, so for instance we're no longer allowed to talk about politics or religion. . . . And veterinarians post pictures, you know, abscesses and nasty wounds and things, so if somebody is having a bad day or something, they don't really want to see that. . . . We try not to talk about medical content, and that one has been the hardest one to enforce. A lot of people's stress comes from . . . you know, we're talking about perfectionists here, they want every single case to go right and when their stress is based on a case not going right, they want advice. But we're not a medical advice board and we don't want to turn into that, and so we try to gently push the conversation away from those topics."

As with most medical professions, stress is rampant and impossible to avoid, but there are also some special aggravations inherent to being a veterinarian. "One of our members put it really perfectly the other day," Carrie told me, "about the emotional roller coaster she has to go on every day at work, where in five minutes she's excited about seeing a new puppy and educating the owners about that, the next five minutes she has to give terminal news, the next five minutes she's counseling someone through a bad decision regarding their pet's treatment, the next five minutes someone's yelling at her about their bill, and it's just . . . you have to have incredible self-boundaries and standards of care in place to handle those dramatic ups and downs. When you consider that a good week for the average vet is fifty hours and a bad week can be closer to ninety, it's a lot to take on, emotionally."

I asked Carrie what sort of conversations tended to take place on the page.

"A lot of it is just people talking to people who understand," she said. "It's the equivalent of going out for a beer with your colleagues after work and letting off steam. It's just that all of our hours are so different and we don't have time to do that in the real world. . . . There's a lot of talk about self-care and a lot of check-ins with people who are having a bad day. Like, veterinarians often skip lunch entirely, and so you know when someone comes on and posts, 'I can't handle my day right now, I'm crying in the bathroom,' it's like 'Hey, have you had any water today, did you eat any lunch, can you go outside for a walk for a minute, how many hours have you slept this week?' These are not things that you need to see a psychiatrist to tell you, but it does start the conversation. We like to use that airplane metaphor, that you have to put your own oxygen mask on first. You cannot be the rock that helps all of your patients if you aren't taking care of yourself first and it's okay and appropriate to draw those boundaries."

I read aloud to Carrie a quote by Viktor Frankl, the author of *Man's Search for Meaning*. It goes like this:

> Unless we treat our patients as three-dimensional beings having not only somatic and psychic dimensions, but a spiritual dimension as well, then the only thing separating us from veterinarians is the clientele.

"I suppose I see what Frankl is getting at," I told her. "But I'm curious to get your thoughts on what you think he's saying here." One of the things differentiating veterinarians from other medical practitioners, in fact, is that the former must deal with both a patient (the animal) and a client (the animal's person). And that can make the job doubly exhausting.

"It's funny," Carrie told me. "We call the person the *client* and the pet the *patient*. The dog does not come into the clinic with a credit card taped to him. He has to come with his person. The animal may be physically ill, but as a consequence of that, their person is going through spiritual and mental anguish. A lot of veterinarians identify as introverts. . . . They're strongly empathetic people drawn to animals but thrust into challenging social situations. Say an older gentleman brings his dog to me, and his dog is quite ill; I can deal with the medical content of that very easily, but the backstory is that this was his wife's dog, and his wife died three months ago and this is his last remaining reminder of her . . . his most important reminder of her. I can do the medicine of the dog easily enough, but with no training in psychology, I have to help counsel him through this moment too, not because I'm really required to, but frankly you're not a very effective veterinarian if you can't thread that human needle as well. It's taxing for anyone, introverted or not."

I asked Carrie if she could think of any cases in which NOMV has, in her opinion, probably saved a suicidal veterinarian's life.

"Hard to know. We surround our members with people who care, calling in welfare checks, getting them to inpatient care—we've done all of that, plenty of times. Other examples include veterinarians who need help, but feel like they cannot afford to walk away from their practice. . . . I mean they own their own veterinary clinic, they have employees, they have to make payroll, and so if they're not at their clinic, the clinic doesn't make money and they feel trapped, which is not an entirely rational thought. That you are so suicidal that you need to go to inpatient care, but you're worried about not going to work tomorrow. But hey, that's okay. We've had the community rally to cover their clinic while they go to inpatient care . . . [or] to take care of the veterinarian's own pets. There was a woman who was losing her vet practice due to some very hard times, and she posted how depressed she was about not being able to give her kids the Christmas that they wanted, so the board bought her kids Christmas presents. You know, it's a community. We've built a community and communities help each other."*

One thing that probably makes NOMV—and other groups like it—so effective is the emphasis it places on the real-world identities of its members. Names and faces are made transparent to everyone in the private network. This way, people genuinely get to know one another as complex individuals (in this case, fellow veterinarians), not just as caricatures with drive-by comments. This structure cultivates true relationships in a sort of online neighborhood. "Don't be an asshole," and you get to stay as a member of a supportive community. Break this rule, and you're out. Sometimes, as in the kinds of situations described by Carrie, these relationships even spill into the offline world, impacting people's careers, families, and overall well-being. When people know that they will be held personally accountable for their words and behavior, altruism flourishes.

Funny how that works.

*In the spring of 2017, Carrie and her moderating team at NOMV were handpicked by Facebook to attend a Chicago summit on its new mission statement on self-harm and suicide. One of only about a hundred groups to be invited, the meeting was spearheaded by the company's founder, Mark Zuckerberg, who was alarmed by the increase of these disturbing behaviors on his company's user platforms. The press-friendly gathering focused on developing a standardized response protocol for group users and moderators, educating network leaders about the relevant applications and tools on Facebook, and providing information about suicide prevention training for gatekeepers of online communities such as NOMV. "We were very honored to be included in the summit," says Carrie. "It made me realize that Facebook takes this issue very seriously. Social media has presented us with a very unique opportunity. There is no other forum where I can type about my day and have, potentially, 10 percent of all American veterinarians respond in a supportive, understanding way."

7

what doesn't die

I'd rather there wasn't an afterlife, really. I'd much rather not be me for thousands of years. Me? Hah!

William Golding

Revelation comes in different forms for different people. A biblical verse. A flash of recognition in a lover's eyes. A Nietzschean proverb. A classical sonata. A child's embrace. Any moment of profundity, really, where time stops and the divine reveals itself, if only for an instant, and the world makes sense.

For me, revelation came in the form of ape knuckles.

When I first met her, Noelle was a six-month-old chimp who'd just been surrendered to a sanctuary in south Florida, where she was to be raised by human caregivers along with half a dozen other orphans like her until they were old enough to live in a more natural captive environment. As one of those lucky caregivers, I volunteered at the sanctuary between classes and spent nights there on the weekends.

Over the next few years, Noelle and I developed, not exactly a daddy-daughter bond, but I suppose something similar to it. One night, as she lay hiccupping on my chest, a belly full of warm formula and drunk with sleepiness, I took her small hairy hand in mine and studied it in detail . . . the crescent-shaped lunulae of her nails, the follicles of coarse black hair at the top of her wrist, the soft pink palm radiating into five delicate digits with arabesque, FBI-worthy fingerprints at each tip, and then the knuckles.

Holding Noelle's hand against my own, the similarity was existentially jarring. Of course, I knew of our shared heritage, that her kind and my kind were each other's "closest living relatives," and I'd observed stunning displays of her intelligence many times. But it was something about her knuckles, those puffy, creased mini-pavilions sculpted over oceans of phalangeal time, set against the backdrop of my own similarly formed material that clarified for me, resoundingly so, not that Noelle was human-like, but that I was animal-like. Or rather, that I was—that I *am*—an animal. Unlike the textbooks I'd been reading in my undergraduate anthropology courses, in which human evolution was always presented as a dry, faraway sort of affair that happened long, long ago in some generic savanna filled with bellowing mammoths, flint, and fire, here was a palpable, thunderous display—nay, a *sacred relic*—of my creatureliness, a profound truth that I could hold and turn over in my hands.

This and similar observations flipped my suburban-hewn worldview on its head.

For a while, even, I struggled to see people as people; like Ionesco and his rhinoceroses, I seemed to be the only one noticing businesspeople and athletes and clerks and professors turning before our very eyes into strangely chattering, adipose-laden, upright apes. It was as if, beneath all the distracting flesh and fashion that led me to assume there's some special mystery to being human, the sturdy skeleton of evolutionary biology—those inalienable facts of life upon which all of our searching pretense and symbolic posturing is built—came poking through for the first time.

I guess many people would be troubled by their cherished assumptions being threatened by such an experience. Yet, I must say, I found it liberating.

Why? Because if humans really are "just" another type of animal, then like every other animal, I, too, was mortal. I felt it in my bones. This is it. This vanishing life.

And then, somehow, as if the words came bubbling up from the same atavistic vein: *Nothing matters. Nothing matters. Nothing matters.*

It became my mantra.

The effect was that, by making me care less, it made me value my life more. That's what the emotion of awe is, really. Stare at an ape's hands long enough, or up at the sky into a vast universe, and suddenly you feel your existence becoming, paradoxically, both more insignificant and more meaningful. I suppose it's the beauty of resigning yourself to the truth of an eternity without you, of being a transient but integral bit of an infinite machine.

Unless you've been there yourself, you've no idea the balm such a thing can offer to a secretly suicidal soul who, his whole life long, had been laboring under the unlikely premise that he had a soul to begin with. What a burden! Without a soul, there's no afterlife; without an afterlife, there's only the theater of the now. Suicide? You'll be dead soon anyway. Eighty years, even a hundred, is the blink of a cosmic eye. Do your worst, and eventually it will be as though you'd never been here at all. Shame has a way of losing its sting once you realize that its venom is man-made.

Even the immortals will perish. "Michelangelo" and "Shakespeare" may some inconceivable day, perhaps in the cataclysmic wake of an epoch equivalent to the span of time that Noelle and I were one and the same species, fall on ears like the pattering of nonsense syllables.

"You might as well live," as the satirist Dorothy Parker—in a slightly different context—penned in her popular short poem "Résumé":

Razors pain you;
Rivers are damp;
Acids stain you;
And drugs cause cramp.
Guns aren't lawful;
Nooses give;
Gas smells awful;
You might as well live.

Nothing matters. Time flies. We're all in this together. Breathe.
There's a spiritual power in nihilism.

There, bathed by the light of a full Miami moon through an open window, Noelle and I lay on a tattered old couch at the sanctuary, two primates, one no less so than the other. I now find the memory scented with jasmine and can hear whippoorwills and the murmur of moth wings against a porch light. It remains among the few times in my life that can, I think, be called a religious experience.

✷

All that said . . . I'd love to see my mother again.

It's been nearly two decades since she died at the age of fifty-four after a protracted battle with ovarian cancer. In the last fifteen years of her life, she'd endured breast cancer (and a double mastectomy), countless rounds of blitzkrieg chemo that made all of her hair fall out and left her constantly exhausted, two of her children being diagnosed with juvenile diabetes, a bitter divorce from my father driving her into a depression that she never quite recovered from, and, finally, the seven-year-long disease that so cruelly took her life.

The specter of death was a taunting presence, always nipping at her heels but forever delaying its fatal pounce. Still, she maintained a quiet stoicism throughout it all, never losing her ability to smile, however dimly, while working full-time in a thankless secretarial job at a local bus company in Fort Lauderdale. She went about her days with a worried brow, but the knot that gathered there was for her children and for their problems, not for herself.

"If you have your health, you've got nothing to worry about," she told me. "Everything else is just a matter of perspective."

It was as if, at some level, perhaps, she knew what went on in my mind. How could I even be thinking about doing away with my life when she was fighting so hard to keep hers?

A few weeks before she died, I sat next to her in the hospital bed that we'd set up in the bedroom of her small townhome. She shared with me that her hospice nurse—a lovely young grim reaper, a Christian poacher of desperate souls—had been proselytizing to her when I wasn't there. She gestured toward the drawer in her nightstand. I opened it to find some tacky evangelical pamphlet, which I crumpled up and threw in the waste bin.

The incident riled me. I called and complained and had the woman promptly replaced. It was the principle of the thing. My mother had been far too polite and physically weak to speak up to this vulture circling over her. But even more upsetting was that I could see how troubling this was to her, because as a "secular Jew," she worried that listening to the nurse's evangelizing was betraying her faith—or at least, what she figured to be her faith.

Like almost every other middle of the road Hebraic person, my mother had no idea what she was supposed to believe about the afterlife. That's not exactly a central point in the Talmud, after all. "There are lots of crazy ideas out there, Jesse," I remember her telling me as she sat propped up against a big pillow a few weeks before she died. Toward the end, she had to stay all day and night like this because of the fluid (ascites) that pooled continuously into her peritoneal cavity, pushing against her lungs and making it difficult for her to breathe. "But there must be *something* after death. How can so many people be wrong?"

It's an important question—how, indeed, can so many people be wrong?—and it's one that I'd argue is even more important, in fact, than Camus's "one truly serious philosophical problem." After all, to live one's life with the absolute scientific conviction that there's no such thing as an everlasting soul, no intangible consciousness freed from the perils and destruction of this mortal world, no self-aware energy that transmigrates to the empyrean plane, should have massive ramifications for our decision making, including when or whether we should take our own lives.

Yet we're done with the question before we can even get started answering it. Because no matter what brand of this-worldly reason a skeptic may pull from their bag of hard-won knowledge in an attempt to explain why any intelligent person would believe in such a thing as Heaven and Hell—wish fulfillment, indoctrination, "out-of-body experiences" caused by altered blood gases, temporal lobe malfunctions, a flailing limbic system, or neurotoxic metabolic reactions invoking vivid hallucinations in dying brains—they're immediately hit in the face with that logical fallacy of never being able to prove the null hypothesis, which in this case, of course, is that there *isn't* an afterlife.

"We just can't know," says the believer. They're right. It's a similar problem, in fact, to that earlier one we encountered in chapter 2 when trying to determine if suicide ever occurs in other animals. "Absence of evidence is not evidence of absence"—that old impenetrable fail-safe of the faithful, credo of the never-convinced and celebrators of mystery.

*

What happens after death?

Perhaps the question itself is wrong. Rather, for those of us willing to concede that the mind *is* what the brain *does*, which is arguably the most fundamental premise in neuroscience, it's simply not a valid question at all.

The famed physicist Stephen Hawking once said, "[I] regard the brain as a computer which will stop working when its components fail. There is no heaven or afterlife for broken-down computers." This hardware-software analogy for the relationship between brain and consciousness isn't without its critics among those who trade in such endless philosophical debates, but to me, at least, the truth in Hawking's reasoning is self-evident. Maintaining one's faith in the afterlife, or even being agnostic on the matter, isn't "acknowledging the limitations of human knowledge"—it's rejecting modern brain science. It's saying that a working brain isn't necessary for thought, which is a preposterous thing to say, really. For materialists, by contrast, nothing could be clearer than the fact that every single subjective bit of us—every quale, quivering emotion, or subliminal percept—is contingent on our brains. There's no arrogance in subscribing to the obvious.

The "hard problem of consciousness," as it's called, is sometimes invoked by those arguing that the mind survives death as a sort of escape clause against materialism. In a nutshell, the problem centers on the relationship between the physical brain and subjective experience; unlike other adaptive bodily systems, which are mechanistic in their functionality, there's no obvious reason why the brain should generate phenomenal states—or in other words, why we have a "theatre of the mind." Couldn't the brain do what it does—learning, processing, responding to the environment—in the same robotic way that, say, the pancreas produces insulin or the heart pumps blood? Why do we need the "feel" of consciousness? It's an intriguing philosophical problem, but does the hard problem have any bearing on the issue of life after death? "I don't think the hard problem has direct implications for an afterlife," the philosopher David Chalmers told me when I asked him this question. Having first identified it—and coined the term—in the mid-1990s, Chalmers's name is synonymous with the hard problem. "Though there has been a small cottage industry of books suggesting it does. It supports the idea that consciousness is irreducible to the physical, but there's a big gap between that and the stronger claim that there is a self or soul that can survive death. . . . [M]y own view is that consciousness is always correlated with physical states and the conscious subject doesn't survive death."

When the brain (a noun) dies, the mind (a verb) no longer occurs. Experience stops; *being* runs dry. It's like Francis Crick, the Nobel Prize–winning co-discoverer of the double-helix structure of DNA, notoriously quipped in his book *The Astonishing Hypothesis*: "You're nothing but a pack of neurons."

Why, then, even for those who acknowledge this codependent relationship between brain and mind, is a belief that the mind survives death so ubiquitous? Ask a believer to describe the difference between their "soul" and their "mind," and if you're not met with a blank stare of utter confusion, you'll likely be witness to a tortured attempt to demarcate between bodily urges and lofty emotions, or perhaps a circuitous appeal to empty placeholder words used interchangeably with soul, such as "energy" and "essence." Whatever it is—and the response may take the form of florid mysticism, theological rhetoric, or even very sincerely felt intuitions—in the end, you will still not have an intelligible answer.

Years ago, I conducted a study in which participants read about a fictitious character who'd found himself on the wrong end of a heartbeat. Here's the abridged version:

> Richard Waverly, a 37-year-old history teacher, was running late for work. . . . It wasn't starting off to be a very good day. Richard felt a little sick and thought he might be coming down with something. . . . He was very thirsty and he thought that some cold orange juice might make him feel better, but he didn't even have enough time to pour himself a glass. . . . While driving, he realized how tired he was. . . . [H]e had stayed up late to read up on the history of the various Amendments to the U.S. Constitution. . . . Richard loved his wife, Martha, but he was angry at her, because she hadn't gotten home until midnight last night and she'd refused to tell him where she'd been. . . . Richard popped a strong breath mint in his mouth and chewed it. As he neared the intersection before the school, he leaned over to close the passenger window, . . . his foot accidentally slipped off the brake and onto the accelerator, and his car went headfirst into a utility pole. . . . A frantic witness called the paramedics, but when they arrived on the scene Richard was already dead. He wasn't wearing his seat belt. They were certain he'd died instantaneously.

Once the participants had read the foregoing passage, I asked them a series of questions about Richard, now that he was dead. "Now that Richard is dead," I asked them, "can he still taste the flavor of the breath mint?" "Does he still remember what he studied last night?" "Is he still thirsty?" "Is he still angry at his wife?" Also, for good measure, and because, well, I'm me: "Can he ever be sexually aroused again?" And so on.

The long and short of it is that participants who believed in an immortal soul tended to carve up their view of dead Richard's mental capacities in terms of his "physical things" and "spiritual things." The former category included more obvious bodily mental states such as thirst, hunger, and horniness; dead Richard didn't have any of these nasty physical things anymore, said the believers. The latter they reserved for more abstract mental states, such as emotions, intentions, and knowledge, all spiritual things of which dead Richard still had in abundance. "Does Richard wish he told his wife he loved her before he died?" I asked them. "Yes," said one participant. "Because he didn't want it to end the way it did."

Most nonbelievers, by contrast, saw the whole mental business closing up shop with Richard's sudden death. "Do you think Richard believes that his wife loved him?" I asked one of them. "No" was the response. "He couldn't believe anything after death because that would involve mental activity."

Yet even the staunchest of skeptics sometimes struggled with the questions. How do we know this? Because in addition to looking at the content of their reasoning (what they said), I also measured the latency of their verbal responses—how long, exactly, it took nonbelievers to say that dead Richard no longer had the capacity for a particular psychological experience. Basically, it took them significantly longer to say that Richard's "spiritual things" were also gone and to get their heads around what "being" dead truly means. In fact, sometimes even *they* got this mixed up. "Does Richard know that he's dead?" I asked one especially outspoken atheist. "Of course he knows he's dead," he told me. "Because there's no afterlife and he sees that now."

When I ran a similar study with young children using a puppet show in which a mouse gets eaten by an alligator, an interesting age pattern emerged in the data. The youngest kids, three- and four-year-olds, were much more likely than older children to say that the dead mouse still retained its psychological capacities after death. That's precisely the opposite trend you'd expect to find if belief in the afterlife were simply a product of culture or something we're taught. The older the child, after all, the more exposure they've had to spirituality and religious teaching. Instead, it tells us that this erroneous belief in the survival of personal consciousness after death is our species' "default stance."

We don't have to learn to believe in the afterlife, in other words, inasmuch as we have to unlearn. Existentialism is cognitively effortful work, which is why it took even those skeptics time to process the fact that dead Richard was now mindless.

Kay Redfield Jamison, in her book *Night Falls Fast: Understanding Suicide*, beautifully captures the role of the physical brain in generating that seemingly infinite galaxy that we know as consciousness.

> Everywhere in the snarl of tissue that is the brain, chemicals whip down fibers, tear across cell divides, and continue pell-mell on their Gordian rounds. One hundred billion individual nerve cells—each reaching out in turn to as many as 200,000 others—diverge, reverberate and converge into a webwork of staggering complexity. This three-pound thicket of grey, with its thousands of distinct cell types and estimated one hundred trillion synapses, somehow pulls out order from chaos, lays down the shivery tracks of memory, gives rise to desire or terror, arranges for sleep, propels movement, imagines a symphony, or shapes a plan to annihilate itself.

To this list I would add, *And envisions a future in the aftermath of its own destruction.*

✷

Maybe it's a glitch in the system.

In the mid-1990s, evolutionary biologists Daniel Povinelli and John Cant put forth an ingenious argument regarding the origins of self-awareness called the "arboreal clambering hypothesis." Once upon a time—about 11 to 5 million years ago, to be more precise—before humans evolved, before australopithecines, before there were anything like chimpanzees or gorillas or orangutans, there was but a single parent species from which all of these later varieties derived. Based on the available fossil evidence, the most likely suspect for this ancestral large hominoid is the rather deliciously named Oreopithecus, an animal that would have been found in what is today's northern Italy during the late Miocene.

Weighing in at about ninety pounds, or about the size of your average sixth grader, and using the tree canopy as its primary means of traveling by slowly clambering from branch to branch, much as its orangutan descendants still do today, the relatively heavy-bodied Oreopithecus faced a huge adaptive problem. A smaller primate could simply swing along on autopilot (or brachiate, as modern gibbons do) without much forethought. Even in the worst-case scenario, a macaque-sized creature would probably survive to scurry away from a ten-meter fall. By contrast, the more substantial Oreopithecus had better choose his next move wisely. After all, one rotten branch or a single tenuous hold while traversing a gap in the canopy would mean almost certain death for this surefooted ape.

This problem of being a big body in the fragile treetops, surmised Povinelli and Cant, was the cognitive spark behind the great fire of what would one day be the human imagination, because solving it required conceptual abilities that allowed the self to go beyond the here and now. By mentally projecting himself into alternative future environments, Oreopithecus could run bodily simulations enabling smart locomotory decisions. *What would happen if I were to put all my weight on this dubious-looking branch? Could I use this pliable vine to sway across that large divide in the tree line? If I stretch my body out just so while reaching to grab that luscious fruit way up there, where should I put my other foot?* That is, this ancestor of ours was able to transcend its immediate sensory experiences and intelligently plan ahead by positioning its body strategically.

Trouble is, once we could project ourselves into the future—running this possibility and that through the sieve of consciousness and envisaging all shape and manner of counterfactual realities—we kind of got stuck there. Imagination is a double-edged sword: on one side are all of our hopes and dreams; on the other, our worst fears. For us apes that eventually came down from the treetops and became *Homo sapiens*, not only do we now anticipate how it will feel to make certain bodily movements (our proprioceptive states), we also imagine the psychological impact of our hypothetical actions and social behaviors. We feel, emotionally, what we'd expect our future selves would feel under those imagined conditions. And when these forecasted emotions spill into the present moment, they influence our decisions. These emotions even include the feelings of a soul without a brain to think them.

If all of that sounds a little obtuse, consider the musings of one Monsieur de La Pérouse, a suicidal character in André Gide's *The*

Counterfeiters. "I stayed a long time with the pistol to my temple," the sad old man tells us.

> My finger was on the trigger. I pressed it a little; but not hard enough. I said to myself: "In another moment I shall press harder and it will go off." I felt the cold of the metal and I said to myself: "In another moment I shall not feel anything. But before that I shall hear a terrible noise." . . . Just think! So near to one's ear! That's the chief thing that prevented me—the fear of the noise. It's absurd, for as soon as one's dead. . . . Yes, but I hope for death as a sleep; and a detonation doesn't send one to sleep—it wakes one up.

Here, the character's suicide is thwarted by an irrational thought that he fully realizes is so, and yet he still can't help but succumb to its powerful influence. But this type of anticipatory reasoning can just as easily add pressure to our trigger fingers.

This ability to imagine ourselves in the future—coupled with the fact that we are, as a species, natural psychologists—means that we tend to have quite lucid projections of other people's reactions to our deaths. We may picture dramatic scenes of mourners weeping at our funerals, of our bullies and antagonists getting their comeuppance and being held accountable, of our friends and family finally appreciating us—all of which, when we're in a suicidal state of social submissiveness, hurt, or anger, may override any other empathic concerns or general hesitations, such as our fear of physical pain.

Vengeance can be a strong driver, too. An excerpt from a suicide note published in the *American Journal of Psychiatry* back in 1959, for instance, portrays what the authors of the article refer to as the "Hostility Directed Outwards Type" of suicide. "I hate you and all of your family and I hope that you never have peace of mind," wrote this person before killing herself. "I hope I haunt this house as long as you live here and I wish you all the bad luck in the world." If only we knew the backstory to that one.

All this is to say that while many suicides simply want to escape unbearable mental anguish and yearn for oblivion, others may be at least partially motivated by the cognitive illusion that they'll be around to experience other people "getting" the depth of their suffering, which they were frustratingly unable to communicate while alive.

Why is it an illusion? Again, if you're of the opinion that a functioning brain is unnecessary for mental experience, that the brain is a nice-enough device but it's mostly just decorative and not essential to personal consciousness, then admittedly the term "illusion" is unfair. But from the viewpoint of scientific materialism, at least, the word does seem to fit.

Many thinkers have pondered this strange psychological phenomenon. "Our life has no end in just the way in which our visual field has no limits," wrote the philosopher Ludwig Wittgenstein, who, incidentally, once said that there wasn't a day that goes by in which he didn't think about killing himself. Likewise, Freud saw the afterlife as a trick of the mind that, in his day, played a role in so many men eagerly enlisting in the First World War. "Our own death is indeed quite unimaginable," wrote Freud, "and whenever we make the attempt to imagine it we can perceive that we really survive as spectators . . . in the unconscious every one of us is convinced of his own immortality." And in his book *The Tragic Sense of Life*, the Spanish intellectual Miguel de Unamuno challenges us to "try to fill your consciousness with the representation of no-consciousness and you will see the impossibility of it. The effort to comprehend it causes the most tormenting dizziness."

Unamuno's frustrations show just how dependent we are on our previous experiences to imagine the future. When we think about unfathomable things to come, we can only draw from our existing reserve of subjective reality. That's all we have at our disposal when it comes to our predictive efforts. In other words, the future is a type of past, and because we've never consciously experienced the absence of consciousness, any such attempt at simulating the so-called state of death will lead us astray. There is no "state" of death. No black*ness*. No nothing*ness*. No painless*ness* or peaceful*ness*. Anything with that suffix implies a perceptible attribute of the environment and, needless to say, requires a perceiving brain.

Even our best analogies—before we were born, dreamless sleep, and general anesthesia—fall completely flat. Yes, we know, theoretically, that we had no conscious experience during these times, but since we didn't actually "undergo" them as periods of non-being, they're utterly useless when we use them to color in the future. "When I try to imagine my own non-existence, I have to imagine that I perceive or know about my non-existence," writes the philosopher Shaun Nichols. "No wonder there's an obstacle!"

In facing our own deaths, therefore, we're just apes clambering to infinity while mistaking it as the next branch in the canopy. Even when we know better, it's difficult not to be ensnared by this problem. When I looked down at my mother's lifeless body in her casket and saw, for the first time that I could remember, no worried knot between her brows, I couldn't help but think to myself that, finally, she was "at peace." It's not that she *wasn't* at peace, either, of course. Rather, she was simply no more.

It's easy to underestimate the significance of this problem, to fail to apply it beyond the setting of, say, a first-year philosophy seminar or a half-baked conversation on your front porch. But again, falling victim to the cognitive illusion of an afterlife—even a nebulous version that lacks halos and other gospel hallmarks but instead has us simply witnessing our own dead bodies, reading our memorials, attending our funerals, or even just being happy, free, or at peace—has real consequences for our decision making.

And that includes deciding to end our lives.

Consider just one of countless tragic cases in which a belief in the survival of consciousness after death can be a dangerously strong trick of the evolved human mind. In Sweetwater, Florida, the parents of thirteen-year-old Maryling Flores forbade her to spend time with her boyfriend, fourteen-year-old Christian Davila, since they thought their daughter was too young to be dating. The two teens killed themselves in a suicide pact by jumping into the nearby Tamiami Canal. Neither of them knew how to swim. "You don't let me see him in this world," Flores wrote in a suicide note to her parents. "So we're going to another place."

But—and there's always a "but" when it comes to suicide prevention—there's also a practical side to fear. Illusion or not, shouldn't we leave people's belief in the afterlife intact for that very reason? In Shakespeare's *Hamlet,* the protagonist reflects on this issue, noting that our dread of a hellish punishment for taking our own lives makes "cowards of us all" when otherwise it should be easy enough to "take arms against a sea of troubles."

One colorful historical account of the anti-suicide function of afterlife beliefs, even those born of vanity, comes from Plutarch, who wrote of a rash of suicides among the young women of Miletus around the year 277 BC. Apparently, teenage girls were killing themselves in a sort of epidemic. "A yearning for death and an insane impulse toward hanging suddenly fell upon all of them," wrote Plutarch. "Arguments and tears of parents and comforting words of friends availed nothing. . . . The malady seemed beyond human help, until, on the advice of a man of sense, an ordinance was proposed that the women who hanged themselves should be carried naked through the market place to their burial. And when this ordinance was passed it not only checked, but stopped completely the young women from killing themselves."*

As we're about to see, the relationship between religion and suicide is complicated, with evidence both for and against the supposition that afterlife beliefs have more than a negligible bearing on people's decision to end their own lives. The statistics clearly reveal that religion is a protective buffer against suicide; in study after study, religious people are found to be significantly less likely to kill themselves—or even to think about killing themselves—than are non-religious people. Scratch beneath the surface of these sociological data, however, and the reason why religion is such an effective deterrent against suicide is made clear—and, surprisingly, it has very little to do with faith or spirituality.

*The reality of suicide is stark. "There is nothing, once you are dead," wrote Thomas Lynch in *The Undertaking,* "that can be done *to you* or *for you* or *with you* or *about you* that will do you any good or harm; that any damage or decency we do accrues to the living, to whom your death happens, if it really happens to anyone. The living have to live with it. You don't. Theirs is the grief or gladness your death brings. Theirs is the loss or gain of it. Theirs is the pain and the pleasure of memory. Theirs is the invoice for services rendered and theirs is the check in the mail for its payment."

✷

Damned if you do.

That was the poet Dante Alighieri's fourteenth-century message for anyone contemplating a self-imposed exile from this earthly realm. In his epic poem *The Inferno*, the souls of those who kill themselves go to a very special part of Hell, taking root in "The Wood of the Suicides." I know it doesn't sound so terrible, the idea of being reincarnated as a tree. The thought of becoming a nice maple in some breezy orchard, or perhaps a melancholy willow serenaded by loons each break of day on the banks of a scenic pond, does have a certain romantic appeal. But that's not what Dante imagined for those so presumptuous as to bring their lives to an end prematurely. Having willfully shed the human form tailor-made for them by God, these souls are condemned to a dark forest in the foul-smelling seventh concentric circle of Hell, their woeful spirits forever entrapped in the newly sprung roots of some random sapling ("wherever fortune flings it," as Dante noted in a clever nod to their rejection of their divinely conceived fate as a mortal) and visited by all manner of evil. Paralyzed by their inanimate nature as sedentary flora, their leaves are grazed upon day and night by legions of odious harpies—clawed, wingéd beasts with fat feathered bellies and screeching human faces. Their branches will snap like crisp little bones forevermore, trampled by the feet of dead heretics who've been sentenced to the same overcrowded forest and who are now being pursued relentlessly by ravenous hounds. Only through this constant suffering, as these poor souls bleed sap from their green injuries, will they be able to "speak"—and by that, the poet meant to lament their decision to kill themselves. Alas, it's too late.

Dante's psychedelic view of Hell was flamboyant, but it aligned neatly with the church's treatment of suicide during the Middle Ages. Back then, *felo de se* (felon of the self) ranked among the most heinous of sins in Christianity. It remains morally wrong across all of the world's major religions today, but there are caveats, as most denominations have at least a vague appreciation of the complex role of mental illness and tend to regard victims of suicide as "not themselves" when they committed the act.*

We'll revisit this ongoing tension faced by spiritual leaders attempting to reconcile archaic theology with sympathy and commonsense in some detail later on. First, though, a whistle-stop tour of how the different major faith traditions have, historically at least, dealt with the problem of suicide. It's not meant to be an exhaustive review, nor one that leans favorably or unfavorably toward religion, per se, but a sort of helpful eavesdropping on the strong moralistic voices that have dominated this age-old matter in religious discourse.

Many people are surprised to learn that the Bible doesn't explicitly mention suicide at all. A few clever theologians, however, found a loophole for this curious omission. In 1637, for instance, a zealous Calvinist minister from Scotland named John Sym reasoned that "when a man, who by nature is most bound to preserve himself, reflects upon himself to destroy himself, the horribleness thereof is so monstrous that we read no Law made against it, as if it were a thing not to be supposed possible." In other words, suicide is so wicked, vile, and unnatural that God considered that its wrongness simply went without saying. Yet in the few instances where it's clear enough that some biblical figure has taken their own life—for instance, Judas (over the shame and regret of betraying Jesus), King Saul (to avoid the humiliation of being captured by his enemies and forced to worship other gods), and Samson (who vengefully killed a bunch of Philistines in the process)—there doesn't seem to be the slightest twinge of judgment for their having done so. Rather, their suicides are simply described matter-of-factly, and the characters' method of death seems mostly tangential to any lessons to be learned from the associated parables.

Only in the early fifth century, when Augustine weighed in on

*According to the Catechism of the Catholic Church, "grave psychological disturbances can diminish the responsibility of the one committing suicide," but the same does not apply to patients with terminal illness: "We are stewards, not owners, of the life God entrusted to us. It is not ours to dispose of."

the subject, was suicide officially classified as a sin. He pointed to a clause in the Bible otherwise known as the Sixth Commandment—"Thou shalt not kill"—saying this rule applied to killing oneself as well as others.* But the suicide debate really heated up in 1485. It was then that Thomas Aquinas sank his teeth into it with his *Summa Theologica*, and with that book's publication, the church's draconian intolerance of suicide became emblematic of the faith.

Aquinas's argument that suicide is a mortal sin—and one of the very worst ones at that—was based on several claims. First, since every living organism naturally desires to preserve its life, suicide goes against nature. (We saw some compelling challenges to that "unnatural" assumption in chapter 3.) Second, Aquinas considered suicide to be, essentially, a form of intellectual theft. Since God designed you and gave purpose to your existence, killing yourself robs Him of any due creative rights.† (Such ownership breeds resentment among the rebellious, however. Hence Dostoyevsky's words of revolt in *Diary of a Writer*, "I condemn that nature which, with such impudent nerve, brought me into being in order to suffer—I condemn it in order

*The German psychiatrist Horst Koch has suggested that Augustine's inspiration for demonizing suicide was sociopolitical as much as it was theological. He was capitalizing on clerical opposition to a fanatical early Christian movement known as Donatism, which had splintered off from the church and was, at the time, thriving in North Africa. Adherents of this movement were being encouraged to martyr themselves as a way of achieving salvation, an abhorrent trend that had been rankling the conservative theologians of Jerusalem.

†There are some remarkable recent findings in the field of cognitive science showing that human beings are "promiscuous teleologists"—we see purpose everywhere, even where it clearly doesn't belong. Ask a six-year-old why the rock you're holding in your hand is pointy, for instance, and he'll tell you that it's so animals won't sit on it, or so it can be used as a weapon, or some such. To say, "it just is," or to appeal to ambient winds whipping against what was once molten substrata over countless millennia, simply isn't in his deck of cards. Now ask your average forty-six-year-old why bad things happen to good people, and, well, you'll get something in the same vein. "To learn an important life lesson," "because that's God's will," "so that we can help others facing similar problems," or some such. Again, it's all due to our species having carved out a special cognitive niche as the animal kingdom's natural psychologists. Pontificating over the *why, why, why* of things comes amazingly easy to us. But delving into the existential *how* of the matter? That's hard work. It's also why atheism is "unnatural": it requires a muzzling of our hyperactive sense of purpose.

to be annihilated with me.") Finally, Aquinas thought suicide was even worse than murder because, unlike killing someone else, you can't repent or confess to the transgression afterward, since you're already, you know, dead. This unatonable quality of suicide rendered it a particularly egregious offense in Aquinas's eyes.

The consequences of demonizing suicides have been well documented. For centuries, not only were Christian burial rights denied to those who'd taken their own lives, but suicide corpses were often mutilated, had stakes driven through their hearts, or were disposed of ignominiously, sometimes being interred at a busy crossroads in the hopes that the constant traffic overhead would keep their vindictive ghosts at bay. (If you're trying to process the physics with that line of reasoning, I'm afraid it's a losing game.) In England, the suicide's property was forfeited to the crown. Even in Aquinas's time, however, you could avoid these indignities if the coroner gave you a medical get out of Hades card conferring the status of *non compos mentis* (not of sound mind). But if you were suicidal between the years 1487 and 1660, good luck with that. Only 1.6 percent of all suicides during this period were adjudicated *non compos mentis*; the rest were *felo de se*. You can see how ulterior motives likely entered the picture with such an arrangement. As the neuropsychiatrist James Harris explains, "The belief that suicide in itself was abhorrent along with the financial interests of the [monarchy] conspired to bias verdicts toward sane self-murder."

Christianity is far from the only religion to view suicide as an unholy act. Most Jewish theologians have similarly interpreted Hebrew scripture as fairly clear on the matter, in that killing oneself, again, falls under the commandment not to commit murder. The Israeli clinical psychologist Yari Gvion points out that "Halakhic law stipulated that Jews who commit suicide are not entitled to a full Jewish burial," and, "indeed, even conducting the rites of mourning such as recital of the Kaddish and sitting Shiva are controversial."

Although Judaism emphasizes this life over the afterlife, Judaic principles still ascribe great spiritual significance to suicide. "When an individual commits suicide," write the scholars Robin Gearing and Dana Lizardi, "the soul has nowhere to go."

> It cannot return to the body, because the body has been destroyed. It cannot be let into any of the soul worlds, because its time has not come. Thus, it is in a state of limbo which is very painful. A person may commit suicide because he wants to escape, but, in reality, the result is a far worse situation.

Muslims too. Like every other major religion, Islam isn't a single monolithic faith but one made up of numerous sects with different ideologies. Nevertheless, these sects are united in their condemnation of suicide. Like Judeo-Christian sacred texts, the Quran does not include unambiguous language on the subject, but there are certainly a few poetic allusions, such as "The lit candle shall burn until daybreak." And the *hadith*, which are considered the most faithful accounts of the prophet Muhammed's actual sayings and actions, include clear denouncements of suicide. According to Abu Hurairah, a companion of Muhammed whose *kunyah* translates as "Father of the Kitten" but whose theology sounds anything but warm and fuzzy, those who fail to heed this warning against wanton self-destruction will have their suicides stuck on auto-repeat and "be eternal inhabitants of Hell":

> He who kills himself with a steel [weapon] will have that [weapon] in his hand and will be thrusting it into his stomach forever and ever; he who kills himself by drinking poison will sip in the fire of Hell, forever and forever. He who kills himself by throwing himself from the top of a mountain will constantly fall in the fire of Hell forever and ever.

With many Middle Eastern countries incorporating the Sharia (Islamic law) as their governing creed, suicide attempts remain a serious criminal offense in places such as Saudi Arabia, Pakistan, and Kuwait. Since suicide is considered a *haram* (forbidden death), those who die this way can expect to be denied burial in Muslim graveyards.

And yes, the militaristic strategy of "suicide terrorism" adopted increasingly over the past few decades by Islamic extremists does ostensibly contradict Muslims' long-standing disdain for suicide. But religion is a protean commodity. In 1997 Osama bin Laden was asked by an American reporter how he and other Al-Qaeda senior officials justified training young recruits to blow themselves up while killing innocent others as a means of warfare. "We believe that no one can take out one breath of our written life as ordained by Allah," bin Laden responded. "We see that getting killed in the cause of Allah is a great cause as wished for by our prophet." In many faiths, not just Islam, a distinction is drawn between suicide and martyrdom. And when you throw in those seventy-two doe-eyed virgins being promised in the Quran to men who die for Jihadist causes, you can see how the Islamic paradise, which is strikingly more sensual than the G-rated Heaven of the New Testament, could be especially appealing to your average young suicide terrorist. Writing in the fifteenth century, the Islamic religious scholar Al-Suyuti said that a martyr could expect something like this in the afterlife:

> Each time we sleep with a houri we find her a virgin. Besides, the penis of the Elected never softens. The erection is eternal; the sensation that you feel each time you make love is utterly delicious and out of this world and were you to experience it in this world you would faint. Each chosen one will marry seventy[-two] houris, besides the women he married on earth, and all will have appetizing vaginas.

Appetizing vaginas aside, you get the idea. When the afterlife is seen as a better alternative to whatever intolerable conditions we face in this world, religious ideas can effectively promote suicidal behavior by exploiting our evolved psychology. There we are again, Oreopithecus's child, glitch and all.

Although Hindu scriptures are ambivalent—or at least ambiguous—about the issue of suicide, for centuries grieving Hindu widows were expected to throw themselves on the burning funeral pyres of their dead husbands in the practice of *sati* (also called *suttee*), a rather horrific ritual of self-immolation outlawed in the early nineteenth century but, in some regions, continuing into modern times. The term *sati* has long been synonymous with "good wife," and the entire concept is basically a big fiery ball of gender inequality. According to one popular line of thought, it was far better for a woman to live out the remainder of her life in "ascetic widowhood" after the death of her husband, but if she and others—mostly others—deemed it unlikely that she'd be able to keep herself from having sex with other men, then diving headfirst into a pit of flames was the second-best option to such a "shameful" existence. The former would ensure her *moksa* (a complete escape from the suffering-fueled cycles of death and rebirth), while the practice of *sati* would only give her temporary accommodation in heaven. How long would a self-sacrificed wife get to stay there, you ask? "A wife who dies in the company of her husband shall remain in heaven as many years as there are hairs on his person." (So, ladies, let that be a lesson to never marry a man with alopecia.) Dutiful wives who die on the pyre were thought to be able to continue caring for and protecting their husbands in the afterlife. They'd presumably also get special supernatural powers, like being transformed into a sort of good-luck goddess and spiritual protector for living members of their families.

And speaking of gender inequality, another example of this blue-skies-after-death effect is an intriguing set of findings reported by Chinese researchers Jie Zhang and Huilan Xu. Unlike most of the evidence in this area, which shows clearly that religion offers a protective function against suicide (more on that shortly), these investigators found that it's quite the opposite in China. There, religion seems to be a risk factor for suicide, especially among religious women. This makes sense in light of the country's popular spiritual beliefs and a systemic prejudice against girls and women. "To some Chinese individuals," explain the authors, "being religious is equivalent to being superstitious, and death is a solution to all the problems and the beginning of a new life. Therefore, it is possible that those who got into extremity are likely to think about starting a new life by ending this miserable one quickly."

In other words, these people are hoping for better luck in their newly reincarnated identities. In a sample of seventy-four attempted suicides, Zhang and Xu found that the strength of one's belief in reincarnation predicted the seriousness of their suicide attempts—and that this was especially true for the female participants. On further questioning, most of these troubled women said that they wanted to be reborn as a male. Given the social premium placed on boys over girls during the decades-long one-child-only policy in China—a controversial population-control legislation that was undoubtedly behind a steep rise in female infanticides and child abandonment—that's hardly surprising.*

Although early Buddhist traditions (especially those that are rooted in the Pāli Canon) are largely secular and supernaturalism-free, later forms of canonical Buddhism prioritize the role of reincarnation and teach that killing yourself will just lead to more problems in the next life, regardless of your gender. People who die by suicide are said to be afflicted by a type of spiritual poison known as *moha*, which translates to "delusion," and anyone who contemplates taking their life as the result of some personal misfortune has lost the plot, basically. First of all, say these Buddhists, death is the ultimate form of suffering and, as such, it's not something that should be embraced before its time. Even worse, suicide is really bad karma. Because you're here to work through your agglutinated bad deeds and unresolved conflicts from your previous lives, killing yourself screws up the whole intricate clockwork process. In fact, it constitutes a karmic demerit and will get you reborn into a lower level the next go-around. Since Buddhism places humans above other animals in a supernatural *scala naturae* sort of way, this means you're destined to be, I don't know, maybe a sea horse or an iguana. Whatever, getting knocked back on the Wheel of Suffering means you're just causing yourself more anguish in the end.

Does the threat of everlasting punishment after death actually deter suicides? In most cases, probably not. Remember that genuinely

*In the late 1980s, the anthropologist Zhou Juhua interviewed fifty-three women from rural Chinese villages. All of these women had recently given birth to girls—a source of marital conflict. Juhua reported that 81 percent of the women wished that they'd given birth to a boy; 100 percent of the women's husbands were "depressed" about the girl and "constantly complained" about not having a son; 60 percent of these men acted in a cold and unfriendly way to their wives; 55 percent verbally abused their wives; 30 percent beat them; and 28 percent of the husbands wanted a divorce.

suicidal people have trouble imagining a worse existence than the one they have now. Hell would be a welcome reprieve.

Ever since Durkheim showed that suicide is more common in Protestant communities than in Catholic communities, researchers have known that it's not merely a religion's harsh doctrine or theology that reduces suicide risk, but something else. After all, suicide is clearly proscribed in both forms of Christianity. Durkheim argued, rather, that the lower suicide rates among Catholics have to do with their high level of societal integration and their degree of moral regulation. Because Protestants, by contrast, are permitted free inquiry, they tend to have fewer shared beliefs and formal rituals, and therefore more permeable groups.

Over the past century, many studies have added support to this "network theory" of religion and suicide. With few exceptions, it's not people's religious beliefs that reduce their suicide risk, but their religious service attendance. In fact, according to the most recent data, if you attend church regularly—as in, every single Sunday morning, and maybe even throw a few weekday evenings in there as well—you're up to *four times* less likely to die by suicide than those who never attend. Just being "spiritual but not religious," or quietly praying to God at your bedside each night? Not so much. "Religious participation and network contacts," summarizes one team of investigators, "[lower the] risk of suicide by encouraging social engagement and support." It's not just the socializing that makes religion a palliative against suicide, though. Otherwise you'd find the same effect with any type of regular communal gathering. It's the socializing *combined with* a purposeful life philosophy.

Just consider the story of Christ. The supernatural stuff aside—virgin births, healing the blind, raising the dead, and the like—it can indeed offer a helpful perspective during times of crisis. "This powerful narrative," argue the religious scholars William Kay and Leslie Francis, "with its recognition of injustice, pain, despair, temptation, opposition, loyalty, betrayal and humiliation, is perpetually represented to the congregation in the readings of Scripture and the preaching of sermons. It becomes a benchmark against which all human experience can be set and by which all human problems can be put in proportion." I agree. Jesus's litany of suffering is precisely of the sort that elicits suicidal feelings in our species. So, for a Christian on the verge of an impetuous act of suicide, I would think that pausing to ask, "What would Jesus do?" can help to calm that gathering storm.

Overall, then, religion is protective against suicide. There's no denying it.

Yet there's a caveat. And it's a big one.

Several studies have shown that devout people who experience religious strain—for instance, because they're convinced that they've committed a sin too major to be forgiven—are actually *more* likely than their nonreligious peers to take their own lives. What would Jesus do after sleeping with his neighbor's wife or squandering his best friend's life savings in a get-rich-quick scam? Well, Jesus probably wouldn't find himself bedeviled by such predicaments in the first place. Thus, feeling alienated from God by their sins, such religious individuals are at a heightened risk of behaviorally accelerating His wrathful sentence.

*

Religion or no, one question we seldom stop to ask is precisely why suicide has been considered morally wrong. Most of us just assume it's the harmfulness of the act, whether to ourselves, to those who care about us, or to God. But according to the psychologist Joshua Rottman, it has more to do with our subconscious feeling that the act somehow "taints" the dead person's soul. And apparently you don't have to be religious—or even believe in such a thing as the soul—to mentally harbor this stigmatizing bias. "Suicide represents an unusual kind of moral violation insofar as the perpetrator of the act is also the victim," Rottman and his colleagues explain in a recent article. "If moral violations are typically considered wrong because of the harm inflicted on third parties, why is suicide so often judged to be immoral?" To get at this question, the authors asked hundreds of mostly nonreligious, liberal American college students to read a fake obituary (which was presented to them as real). Details about the deaths were scant, but the decedent was said to have died either by suicide or murder. With this exception of the manner of death, the obituary was identical:

> Louise Parker, who was 68 years old, died on January 11, 2008, due to [suicide/homicide]. Louise had always been very close with her siblings, and had recently spent the holidays with all five of them. Her brother Roger wrote, "Louise was a terrific sister. She was a joy to be around, and always knew how to make a person laugh. Her charm and energy were contagious and appreciated by everyone who met her. Louise couldn't go anywhere without running into people she knew. I've been truly lucky to have spent so many quality years with her." Louise is survived by her brothers, Mark and Roger, and three sisters: Geraldine, Karen, and Theresa. Her memory will live on in the hearts of many.

The participants in the study were then asked how they felt about the death. The researchers wanted to know the degree to which people thought it was morally wrong for Louise to kill herself (in the suicide condition) or for someone to have killed her (in the murder condition). Importantly, they were also interested in whether or not people's responses to this question correlated in any way with their judgments about the perceived *harmfulness* of her death (e.g., "Did Louise's suicide/homicide cause harm?") and her *purity* (e.g., "Was the purity of Louise's soul tainted as a result of her suicide/homicide?").

As you'd expect, the participants rated both types of death as very harmful. But it was only when Louise died by suicide that they saw her soul as defiled by an impure action. In other words, her soul remained clean and unsullied if she was murdered, but not if she killed herself. And in spite of their post hoc rationalizations about her suicide being wrong because of the harm it caused (to her siblings, for instance), the subjects' moral judgments were, in fact, guided more by the vague belief that Louise's decision had, somehow, contaminated her immaterial soul. That's to say, there was no link between the participants' ratings of harmfulness and wrongfulness; the moral correlation solely involved her dirtying her spirit.

Related to Rottman's findings, it would seem, is the common belief that some intangible essence of a suicide seeps into the very architecture of the building in which it occurs. In O. Henry's spooky short story "The Furnished Room," a man rents a room at a dingy boardinghouse while he searches the nearby theaters for the woman he loves, a young singer who has abruptly left him for a stage career. When he inquires if his darling beloved has been there, the savvy housekeeper lies to him, saying no such aspiring singer has ever stayed at their facility. Eventually, however, we learn that the missing woman, distraught over her lack of money and fame, in fact recently took her own life in the very room—indeed, the very bed—where our heartbroken hero is staying (which accounts for his sixth sense that she's somehow nearby). "It's the money we get for the rooms that keeps us alive," we overhear the old landlady saying to her friend, justifying her deceiving the man. "There are many people who wouldn't take a room like that if they knew. If you told them that someone had died in the bed, and died by their own hand, they wouldn't enter the room."

Real life is even stranger than fiction when it comes to real estate disclosure laws. Although it's relatively easy to come up with a list of logical reasons why an agent should have to inform prospective buyers about, say, a homicide that occurred at a house that's just gone on the market (perhaps the killer was never caught, it's indicative of general crime in the area, something about the home attracts murderers or thieves, and so on), it's not entirely self-evident why they should be obligated to do so in the case of suicide. Nevertheless, many jurisdictions worldwide classify such homes and apartments as "stigmatized properties," and realtors are often legally required to disclose this sensitive information.

In 2014 a couple in Auckland, New Zealand, was having buyers' remorse, feeling that their new home was "dark and felt sad and depressing." After sharing these niggling feelings with a neighbor, the couple learned that the previous owner had hanged himself in the garage the year before. They henceforth filed a lawsuit against the real estate company for failing to share this detail with them, because when they relisted the home with its macabre past revealed, they now found it impossible to unload it without taking a huge loss.*

How would *you* feel about living in such a house? If you're not particularly superstitious but are still creeped out by the thought of calling the place home, how do you explain your knee-jerk discomfort? "It's like the jersey effect," explained Joe Nidd, co-owner of a local real estate company here in Dunedin. I'd phoned him to ask why he thought so many buyers are uncomfortable with the idea of purchasing a home where there was a suicide. "I heard about it on this podcast called *The Skeptics' Guide to the Universe*," he went on. "It's not rational, but people just feel like they're going to be exposed to something bad, I guess."

Back in the 1990s, the psychologist Paul Rozin conducted a now-classic experiment in which people were asked about their willingness to don some object of clothing, such as a jersey or sweater, that had once been worn by a serial killer, sex offender, or some other kind of social pariah. Most participants abjectly refused to do so—even after they'd been assured that the garment had been thoroughly laundered. But when it came to explaining why they felt so unnerved by the idea of wearing these (completely sterile) clothes, they found themselves stumped. Jonathan Haidt, who is one of Rozin's colleagues, has a special term for this inability to articulate why we feel the way we do about taboos. He calls it "moral dumbfounding." Basically, we can't always unpack verbally why certain behaviors, such as, for instance, wearing a rapist's sterilized eyeglasses or taking a bubble bath in a clean tub in which someone once died by slitting their wrists, are wrong. Logic falls flat.

It just *feels* wrong.

*Ultimately, they lost on appeal after the case went to the High Court. In the United States, disclosure laws vary dramatically by jurisdiction. Some states, such as California and New York, require that real estate agents must provide such death-related details even if unsolicited; other states, such as Utah, mandate that brokers share this information only if asked directly by buyers; and still other states, such as Ohio and Tennessee, advise nondisclosure of these morbid facts even when the agent or seller is asked outright.

It's almost as if, as Joe pointed out to me, we fear we'll be contaminated.

"But by what? Evil?" I asked him.

"Look, it's not about what you or I believe," he said. "It's about whether the reputation of the property is going to put off other buyers and affect the value of the home. Even if you knock down the house where the incident occurred and rebuild on the site, realtors have an ethical duty to let the customer know there's been a murder or suicide on the grounds. But an old man dying in his sleep in an armchair, that sort of thing, it's not really important information." In other words, we might not be willing to call it a tainted soul, but many of us do seem to worry, at some level, about a contagious bad essence remaining after a suicide.

This is a universal phenomenon, too. If a person dies by suicide in some parts of Ghana even today, the body is removed through a window or a special hole cut into the wall so that the doorway isn't desecrated permanently. If one hangs oneself, the tree is felled and burned. Closer to home, so-called mediums and paranormal investigators do frequent house calls to places where the owners are being rudely kept awake at night by ghosts. In Japan, where dense populations in urban centers and a growing sense of isolation and disconnectedness have given rise to a preponderance of "lonely death apartments"—which are properties in which an occupant who lived alone has died—superstitious beliefs about lingering negative energies are rampant. These *jiko bukken* (black properties) are difficult to rent out and can often be obtained for bargain prices; in fact, some clever landlords have recently resorted to rebranding *jiko bukken* as opportunities for single people to live with friendly ghosts.*

Of course, when we actually know the dead, this phenomenon isn't always scary. For many people, living in the house where a loved

*This isn't a new trend in Japan. In 1927 the writer Ryūnosuke Akutagawa left a suicide note in which he added: "I feel envious of a bourgeois who can afford to own an extra house or a villa in which to commit suicide." It's also one of the reasons why notorious hot spots such as Aokigahara at Mount Fuji, better known as the Suicide Forest, are so popular.

one died, even by suicide, can be comforting. "I felt guilty [moving out of] the apartment because it felt like I was leaving him," my former PhD student Claire White-Kravette shared with me recently. Originally from Belfast, she's now a faculty member in religious studies at California State University, Northridge. She's also the type of person that you can talk to about anything.

The day after Halloween in 2015, Claire came home from the gym to find that "Aaron," her best friend from Ireland who'd been staying with her for the past two weeks, had hanged himself in the shower of the bathroom in her apartment. The song "The Bitter End" by Placebo was playing on repeat on the computer.

"It's a fucking horrible song, Jesse. Have you heard it?"

"I don't think so," I said.

"It's about suicide. The guy who wrote it is so depressed, if you listen to the lyrics."

"So, Aaron played that on purpose, for you, when you found him?"

"Yup."

She's since gotten divorced and is now remarried, but I'd met Aaron at Claire's wedding about eight years earlier. A lively gay man with a wry smile and about the same age as me, I found him to be warm and engaging, happy to skip the small talk and keen to delve into heady discussion. Since then, he'd apparently run into some hard times. Having been laid off from his "dream job" as an occupational psychologist in San Diego, Aaron found himself without a green card and was forced to return home to a dreary, isolated farming village near Donegal. There, he'd taken over the daily care of his mother, who was suffering from advanced Alzheimer's disease. Overcome with stress, loneliness, and a sense of failure, he started drinking heavily.

"How long did you stay in the apartment afterward?" I asked Claire.

"Nine months," she said. "I was alone, just divorced, and everyone thought I was crazy. They kept saying move. But he was my best friend when he was alive and, I know it doesn't make sense, but he was still my best friend. Why would I want to leave him?"

"It must have been incredibly difficult, having to use the shower every day and—"

"Do you know what? It was actually reassuring. When I used the shower, it was a bit weird. I kept thinking something was going to show up. I kept thinking there was going to be a note or a sign, or I kept expecting to see him in the middle of the night. And the shower thing, I put an angel—I'm not personally religious at all—but I did put an angel above the shower. Every time I'd go to shower, I just felt a little bit . . . I mean I even . . . I reenacted it. Like four times. The biggest drive for me was to understand how this happened. I literally put a towel around my neck, just as he'd done. I know that sounds nuts."

"Not at all," I said.

"I wanted to see, what was the last thing he saw? What was the last thing he heard? I closed the door. I put the fan on. Had the water running. What was it like in there? It was a tiny space. I knew he was claustrophobic. You know, I saw handprints on the top of the shower, in the dust. There were still fingerprints on the door frame where he'd grabbed."

"You said you were looking for signs. Did the atmosphere seem to change?"

"Yeah, in the bedroom. Not in the bathroom. In the bedroom where he slept. There was a distinct smell that he had. Probably wine . . . ha! It was still there and it was eerie. You know when you go in a room and someone's smell is there? The first time I cleaned the room, it was horrible. I didn't want to clean it because it felt like I was cleaning away . . ."

"Him?" I asked her.

"Yes!" said Claire. "I used to wake up in the middle of the night and I could feel . . . I'm kind of embarrassed even to talk about this. I felt a presence. And by presence I don't mean a physical presence, like his body standing there. I just sensed an energy. I don't know. I just felt like somebody was there, and I knew it was him. Upon reflection, I don't think that he was there. But I had the experience of him being there, if that makes sense."

"Yes," I said. "It does."

"It had an effect on me. I felt comforted. It felt familiar. But I didn't, but I don't . . ." Claire sighed. "I don't know. I believe the experience felt real, I guess that's what I'd say."

"Would it be different if it were, say, you moving into the apartment and you didn't know the person who'd killed themselves and you found out—"

"Oh God, I'd be completely and utterly freaked out!"

✷

Recently, my research colleagues Emma Curtin and Jonathan Jong and I attempted to get to the bottom of why so many people are dismayed at the idea of inhabiting a physical space in which a stranger has died, whether the life has been lost to suicide or some other cause. In this study published in the *Journal of Death and Dying*, we decided to use the scenario of staying in a hotel room, a common setting for suicides who are wishing to avoid, among other things, "psychologically impacting" their own homes.*

First, participants read a brief vignette. "Imagine that you are traveling on a long-distance road trip in the United States," went the story,

> and, given that it's getting late, you have decided to pull over to the nearest hotel to get some much-needed sleep. On checking in, the manager at the desk notifies you that you are the 100th guest that week and, as such, they are pleased to inform you that you have won a free two-night stay. There are only two rooms available, however. One of these rooms is an average, standard guest room that typically goes for $105 per night. The other room, by contrast, is a more luxurious, upmarket suite that is typically priced at $215 per night.

*We figured that hotel rooms were a better option for a controlled study than houses, because this way we could avoid participants' more "rational" concerns about the resale value of a house, such as people avoiding a home purchase not because they might have irrational beliefs about the property themselves, but out of consideration that *other* future buyers would be superstitious.

Alongside this information were photos of the rooms in question, one being much nicer than the other. The one had fancy furniture, great views, high-end electronics, and similar plush trappings, and the other was just your standard Holiday Inn variety setup. You'd go for the luxury suite, right? I mean, who wouldn't? Oh, but there's just one catch:

> Before making your decision about which room you want, the hotel manager tells you that they feel it necessary to inform you that the last guest to stay in the luxury suite died in the bed.

How did this previous occupant die, you ask?

That was a key variable in the study. Participants were randomly assigned to one of three different "death conditions." Those in the *natural* condition read that he died "from an allergic reaction to prescribed drugs"; those in the *homicide* condition learned that he died "after being deliberately poisoned with drugs by his spouse"; and people in the *suicide* group were told that "he died in the bed after purposefully overdosing with drugs." So, it was drug-related in all three cases—this way, we could minimize participants' quite rational concerns over things like blood spatter and infectious biohazards, and focus instead on their irrational aversions—with the only difference between them being the decedent's intent. There was also a control group of participants who weren't told about any death.

Not many people in the control group went for the more boring standard room, needless to say. As expected, they selected the fancy room. But it was a very different story for those who'd been told about a recent death in that space. No matter how they thought the previous occupant had died, most people in the study were pretty weirded out at the idea of an overnight stay in a room where someone had recently lost their life. In fact, when given the choice between the two free rooms, 66 percent of participants opted for the crappier (and yet death-free) accommodation. Moreover, when asked how much they thought the rooms were "really" worth, those participants who were told about the death estimated the luxury suite to be worth substantially less money (≈ −26 percent) than did control participants who were asked to price the very same room.

And interestingly, the people in the suicide condition were the most likely of all to explicitly mention the death as the reason for their devaluation of the luxury suite. "I would never pay more than $100 for a room in which someone died," for example, or "the fact that someone died in the room is just creepy." Those in the natural death or homicide conditions, by contrast, tended to deny any superstitious beliefs and tried to offer more rational-sounding explanations, such as "Hotel rooms are always overpriced," or "I just estimated the cost." If most of us are operating, implicitly or explicitly, with the assumption that suicide is morally wrong, this difficult-to-articulate feeling of some intangible *essence* of a suicide permeating the physical world can be seen as a particularly sticky cognitive bias that's hard to scrub from our reasoning even when we know it's superstitious nonsense.

In a more recent study, my colleagues and I cut right to the heart of the matter, so to speak, and surveyed people on their feelings about receiving an organ transplant from a donor who died by suicide versus other causes. "Imagine you are in need of an organ transplant," we started.

> Your doctor has found three possible matches for you from three different healthy young donors, all of similar age. The doctor assures you that any one of them would be an equally good match for you and that the donors differ only in the way that the person died. One donor died in an accident; one died by homicide; and one died by suicide.

After they'd read this information, we had participants rank order the three prospective donors by whose organ they'd most and least prefer to receive as a transplant. Now, I know what you're thinking, all you hyperrational folks out there. *It doesn't matter*. If you were that desperate, you'd just be grateful to have a donor—any donor. Plenty of participants in our study told us the exact same thing. "I would hope that if I were unlucky enough to need a donated organ,

I would not be concerned with the circumstances of the donor's death," one person said. "I wouldn't care at all. Freshest and healthiest organ. That's what I'd want," said another. "Personally, it really wouldn't matter how the person died," said another. "If there was a match and it was going to save my life, then it won't matter who it came from." And so on. And yet, just like almost everyone else in the study, each of these perfectly rational, perfectly non-superstitious people ranked the suicide donor dead last. That is to say, there was a massive disconnect between what people told us (and probably also told themselves) and how they actually felt about having a suicide donor's organ inside of them.

In a follow-up experiment, we gave participants details about the donor—"he was a generous and kind person with a good sense of humor . . . enjoyed eating spicy foods and his favorite color was green . . . but also had a darker side [and] could be impulsive, moody and argumentative," and so on. Compared to participants who read such idiosyncratic descriptions about a donor who'd been murdered or died in an accident, those who read about a donor who'd taken his own life were significantly more likely to reason that his quirky habits and aspects of his personality would also be transmitted along with his vital organ.* "I'm sure this has no scientific basis," one man reflected on his own train of thought. "But I would be afraid that if I accepted the organ of the person who died by suicide, that could somehow cause me to suffer from depression and possibly contemplate suicide."†

✷

"All right," I said. "Here's kind of a big juicy question for you."

Sitting at the table with me were a Progressive Jewish rabbi, an Anglican priest, a Thai Buddhist abbot, and a Baptist minister. I'd

*"Compared to how you were before receiving the transplant," we asked them, "please rate the extent to which you think the following unique characteristics of the decedent are likely to affect your personality and/or behaviors now that their heart is keeping you alive."

†Another expressed it in a more explicitly superstitious way: "dying by suicide seems like bad mojo on the organ."

gathered this motley spiritual crew at my university office for a sort of informal conversation to help me pick apart the complex relationship between suicide and religion. It had already been a fruitful exercise. But this question I was about to pose to them, above anything else we discussed that sunny afternoon, was the one I'd been champing at the bit to ask.

"As religious leaders, I'd like to know your answer to this, just if, you know, push came to shove. Let's take a 'stereotypical' suicide of, say, a middle-aged man who's experiencing financial difficulties, relationship problems—maybe his wife just left him—but otherwise in good physical condition. He's not suffering from any sort of terminal illness or bodily pain. Finds himself on a bridge, jumps, and ends his life. So that's the sort of scenario I've got in mind when I'm asking this question. Are you ready? Here it is: What becomes of that person after death? Is there a future for that person . . . and if so, what *is* that future?"

"Future?" said the abbot. "In Buddhism, we believe that there is an unfortunate place for him after death. Because life is a precious thing; it's like a treasure. When you stay alive, you have more time to do something better or to improve or to fix something."

"So, this person who died by suicide," I replied, "he disrupted what he was supposed to do? Is that what you're saying, or am I misinterpreting you?"

"It will take many lifetimes for him to sort out. It's like a gene or chromosome comes with your spirit or your mind and then you come back through rebirth, it will happen to you again and again until you find a solution. Every time we have a problem, we have to see or talk to someone. Suicide is not the solution."

"Okay, Rabbi," I said. "I'll put that same question to you: What happens to this person after he kills himself?"

"In progressive Judaism," said the rabbi, "we tend to give the person the benefit of the doubt. Even if this person left a note and was clearly planning on killing himself, between throwing himself off the bridge and hitting the ground, they may have changed their mind. They may have repented on their way down, before they died."

"I see," I said. "I just want to clarify, though. Because in this case—

jumping—that might be true. But a gunshot? I mean that wouldn't leave much time to repent, would it?"

"Well, they might have changed their mind. They didn't have guns two thousand years ago. But in any event, Judaism will say that we either have a place in the world to come, go to Heaven, whatever that means, or we don't. We don't know much about the world to come because we haven't been there and, really, the best we can do in terms of preparing for it is live this life as best we can, achieve our potential as best we can. So, I think that's a different emphasis from some other faiths." The rabbi paused, weighing his next words carefully. "Having said that," he went on, "I don't really believe in life after death. As far as I'm concerned, when we die we sleep forever, and I expect he'll sleep forever too."

"Interesting," I said. "So that's your personal view, then?"

"If I'm wrong, I have to live with the consequences," said the rabbi. "Die with the consequences. My feeling, and I've seen many people die, or been with them, you know, in the aftermath, and I would like to think that it's just like a deep, permanent sleep. And I'm not persuaded by the other options, either. We really don't know. But I would tell the man's family that his soul has returned to the shelter of God's wing, as much as anywhere else. I wouldn't deny anybody that perpetual care and love and relationship, whatever there is."

"What do you think about this question?" I turned to the priest. "What would you say is the likely fate of this man's soul?"

"I think I need to make a distinction between what my religion would have taught at one point and what I actually believe," said the priest. "I'm pretty sure that my tradition would have said that the person is cut off from God. However, I can't bring myself to believe that. I do believe in life after death, and I do believe there's a judgment. But I'm agnostic about what that person would face."

"I see," I replied.

"I would like to believe," the priest continued, "that they would face the mercy of God rather than a sort of . . ." He paused, shifting in his seat. "Well, the mercy of God rather than the justice of God, as it were. But again, I do stress, that's what I would like to believe, and I think I have reasonable grounds in other parts of my tradition for saying that, in the scriptures and the character of God, that I think is revealed in the scriptures, and particularly in Jesus Christ. But I'm aware that's, to some extent, in tension with my tradition."

"Do I believe in life after death?" said the minister. "Yes. I often find myself finishing a funeral by saying, 'And we commend so-and-so's spirit to the God who gave it.' I don't make judgments at these funerals."

"So, you've done funerals of people who've died by suicide?" I asked him.

"I've done maybe ten, maybe twelve over the years. I've done a few secular funerals, in the sense that there is no evidence of a faith in the person who died. But all their friends who get up and speak and say, 'I'll see you later,' I've reflected on this quite often—is it a safety thing? Is it a safety saying, or do they actually believe they're going to see the person again? I remember this one where there was a kid killed in a car accident . . . well, he probably killed himself, actually, because it seemed to be a deliberate attempt and fortunately he didn't take anyone else out with him. But there were about a half a dozen of his buddies who got up and in their own way they all said, 'Yeah, we will see you again.' I find that fascinating."

"So do I," I said. "It could be that's just the language people use when . . ."

"No—," interjected the minister. "They were being sincere, in my view."

We sat for a moment in silence, pondering the weighty question hanging uncomfortably in the air. I joined these thoughtful men in the enigma.

Still, I couldn't help but look at all those knuckles on the table.

Five primates, none of us more so than the others.

8

gray matter

> Disgust with our own existence, weariness of ourselves is a malady which is likewise a cause of suicide. The remedy is a little exercise, music, hunting, the play, or an agreeable woman. The man who, in a fit of melancholy, kills himself today, would have wished to live had he waited a week.
>
> Voltaire, *Philosophical Dictionary* (1764)

Early on in the course of writing this book, I was advised to never, ever use that dreadful phrase "committed suicide."

Such a thing implies, I was told, a criminal act: the commission of a crime. A cruel linguistic artifact from those judgment-heavy days of *felo de se*, when suicide really was a prosecutable offense (either against the person himself in the aftermath of a botched attempt, or as a financial punishment levied by the government against the victim's family through seizure of goods and property), today the expression "committed suicide" still carries with it an air heavy with legalistic sinfulness. Not only does it stigmatize the mental illness, whatever it is, that's thought to underlie the act, but it also pollutes the reputation of the dead, as it presupposes the individual's failure to rein in the responsible defects of their character, such as selfishness, a violent nature, or cowardice.

"Do we ever say that someone 'committed cancer' or 'committed heart failure,' even when they may have lived lifestyles that contributed to such diseases?" asks the Australian suicide prevention specialist Susan Beaton.

And speaking of failures, we should probably also do away with, say many experts, our peculiar habit of labeling sub-lethal incidents of self-destruction as "failed suicide attempts," as well as those that end tragically in death as "successful suicides."

"My 'failed attempts,'" said one patient, "made me feel just like that, a failure. Once when I drove my car off the road, in the ER the nurse asked me if I had taken my seat belt off. When I replied no, she rolled her eyes and shook her head."

"As health professionals," Beaton explains, "our goal must be to reduce the taboo surrounding suicide. To do so, we must update our language."

What are the appropriate ways to talk about this dismal matter then, you ask? The phrase "died by suicide" is preferable, suggest some mental health advocates, to the subtly larcenous "took their own life." Or, in what is an unspeakable affront to my own ears, even if it's not to the English language, the verb *suicided* is also acceptable.*

In any event, I've learned along the way that the particular terms used in suicide prevention discourse separate the dilettantes from the professionals. I do think that shifting the vernacular in these ways has admirable aims. Just between us, though, I struggle to muster up much passion for these righteous semantic feuds. They feel a little overzealous to me; and a politically correct nomenclature always risks slipping into pedantry.

Although words are powerful and they can indeed lead a recalcitrant culture by a leash into greater understanding, there's also the faint pretext here that one is actually doing something meaningful to combat the problem of suicide simply by adopting the approved industry terms. It's a fine discussion to have, this debate over language, but we need to pick our battles, and this one feels rather trivial to me in the overall war.

"Not sure if you've heard," my friend Jack wrote to me, "that my

*"How old was Robin Williams when he suicided?" someone asked me recently. This usage is so bizarre that it took me a moment to parse the sentence. He was sixty-three, incidentally.

father hung himself last August. I still can't believe it, he actually committed suicide."

"Oh, Jack," I responded, as any human being would. "I'm terribly sorry." Just imagine if my instinctive reply was instead: "Actually, Jack, it's *hanged*. You mean to say that your father *hanged* himself. Also, 'committed suicide'? What is this, the Middle Ages?"

The phrase "committed suicide" is obsolete, yes. It's suggestive of sin and crime and drags with it the weight of a thousand dark years. Still, it's the intent of the speaker that matters, and no matter how tactfully your lesson is rendered, pointing these facts out to a friend who just shared the worst news of his life with you is a pretty dubious contribution, in the scheme of things. Besides, the topic is already difficult enough to talk about.

I'm all for elevating the public discourse on suicide, of freeing it from the ecclesiastical cells and punishing court dockets where it festered for so long. But etiquette is key, and I fear if it's not done with great sensitivity, such a supercilious language campaign will only introduce a new land mine that will make many people not dare to discuss the topic because, on top of everything else, now they're keenly afraid of making a faux pas in their word choices.

Jack, for his part, couldn't care less about such things. *Hung* or *hanged*, *committed* or *died by suicide*, his beloved dad is dead. What's more important to Jack, a recently minted PhD in philosophy, is coming to terms with a devastating suicide that, from his point of view, was also not without a certain rationality.

"He was a logical thinker," Jack told me of his father, Ross, a retired mathematics professor. "And understanding the way he thought, I think he made a logical, even if desperate, decision. He wasn't 'mentally ill,' which most people mean as an inability to reason . . . or basically, *insanity*."

Around three o'clock in the morning on August 25, 2015, sixty-five-year-old Ross took a few plentiful swigs of vodka to steel himself for what he was about to do, tightened the noose around his neck that he'd meticulously prepared, and hanged himself in the garage of his large country home, much of which he'd built lovingly with his own hands.

"Being an applied mathematician," Jack said, "he was able to calculate weight and measure the drop and that kind of thing, to make sure it was done properly. People can get it wrong, and he did everything with extreme precision. His feet came to rest just a few inches from the floor."

A few months later, after the materials from the scene had been processed, an ill-informed police officer notified Jack that they'd be sending him, as his father's legal representative, all of the collected items: letters, names and addresses of intended recipients of the suicide notes, the cell phone . . . even the noose itself.

"It beggars belief, doesn't it?" Jack remarked about the noose. "But when I got to the police station, they reneged on that and kept it. I was actually a little disappointed I didn't get it. I wanted to chuck it in a river. Or, I don't know . . . Dad had actually left me a set of instructions called 'Possessions and Final Instructions.' His wife's lawyer strenuously dismissed these instructions as 'not a legal document' . . . as if his words were not clear on his intentions. So, I also pictured myself throwing that noose on a courtroom floor, saying, 'That's not a legal document either, but it makes it pretty clear what his intentions were!' "

The prospect of handling that intimate object of his father's demise was just one of many surreal experiences Jack has had to face. Like so many others who've lost a loved one to suicide, he soon found himself doing the gritty detective work . . . comparing the alternate versions of suicide notes that his father had typed out on his laptop—some angry, some more composed—mulling over possible double meanings in emails and text messages ("I'm literally at the end of my rope" is one such line that still haunts Jack), interrogating everyone who saw and spoke to his father those last few days, even retrieving deleted messages from his phone, all in an attempt to piece it together, to somehow comprehend how it had all gone so terribly wrong for the man he so admired.

Jack's sleuthing was done alongside a formal coroner's report, a standard course of action for suicides that aims to assess whether a police inquest is warranted. Also, the regional mental health-care team that had treated Ross prior to the "serious adverse incident" (aka the suicide) was audited to ensure that everything was indeed done by the book. Taken together, a picture began to emerge of a brilliant but complicated man whose pride had been deeply wounded upon discovering that his new wife lied to him about a previous relationship.

"Dad was a passionate person. He was sensitive, but also had a temper," Jack told me. "I'd been on the receiving end of his wrath before."

This time, though, it was his wife who bore the brunt of it. And this time, words swelled into a blow and her fleeing the scene. A passing driver was flagged down in front of the couple's home. The police were called. She had a fat lip. Now a trial was pending. The court mandated that Ross and his wife be separated until the hearing. She went to stay with family abroad. He was left to stew alone in that big country house.

"When I heard what had happened, the situation he was in, I immediately thought 'That's it,' said Jack, who lives in another city, far away from his father's place. "Because I know what he's thinking . . . he knows he can't forgive her for it . . . even if he did, he won't be able to live with all that had happened. It was a nasty row. Anyway, I actually thought, 'That's it, there's no way he's coming out of that.' "

Jack asked his dad outright if he planned on killing himself.

"Dad," he said, "you're not thinking of doing anything stupid, are you?"

And Ross outright denied it. "He sent me a text saying, 'I have no intention of taking my own life at all. I'm not in any danger. Stop interfering.' They're the parent, you know," Jack explained to me. "Even though you're thirty-nine years old, or whatever, you're made to feel like you're the little kid and don't know any better. So, you sort of think, 'Nah, he wouldn't do that, because he says he's not going to do that.' "

But Jack still had a bad feeling.

"I was always checking in with him during this time, and he was just very unresponsive. You'd kind of have to jolt him into replying. And when he did, it would just be very short. 'How are you doing, Dad? How are you feeling?' I'd text him. And I'd get a single word back—'*dark*.' "

When his father stopped responding altogether, Jack and his aunt (Ross's sister in Canada) grew alarmed and called the authorities for a welfare check. When the police showed up at his house, Ross was his normally affable, professorial self; he agreed to let them take him to the hospital for an assessment. "He was compliant. He said, 'Yeah, no problem, whatever you need to do,' and they took him straight to the psych ward for an evaluation."

This was, mind you, just one week before Ross took his own life.

Jack showed me the "Serious Adverse Incident Report" for his dad's case. It included details from the consultation with the resident psychiatrist that fateful night at the hospital. The document illustrates just how challenging it can be, even for trained professionals, to accurately assess a person's suicide risk.* "The subject was relaxed in manner and spontaneous in conversation," reads the report.

> He spoke about the devastation of his wife leaving and the humiliation of the pending court case but he was positive about the future. . . . Presented as casually dressed and well-kempt. His face was flushed, there was good eye contact and rapport was established. He was a good historian. . . . It was felt by the team that the subject did not suffer from a mental illness but was experiencing a reaction to a situational stressor. . . . He was very articulate and talking in detail about his life. The Consultant Psychiatrist confirmed that the subject was sad and heartbroken but not clinically depressed. The subject readily acknowledged that everything would be okay if the charges were dropped and his wife agreed to reconcile, i.e., it was a situational remedy, not mental health treatment, which was required.

The psychiatric nurse advised that Ross be admitted overnight. But around 1 a.m., Ross phoned Jack to say that he wanted to go home. Jack cautiously agreed; he felt that keeping his father at the hospital against his will would only make matters worse, adding yet another "situational stressor" into an already volatile mix. So, Ross

*Ross's is a very different story from Vic McLeod's, but like her journal, it highlights many of the psychological mechanisms we've seen throughout our journey.

went home, as he wished, with prescriptions for diazepam (for his anxiety) and zoplicone (a sleep aid) in his pocket.

The psychologist Matthew Nock and his colleagues point out that simply asking someone if they're suicidal isn't usually very effective. "This approach is limited," write the authors, "by the fact that people often do not know their own minds and is especially problematic in measuring suicidal thoughts because people often are motivated to deny or conceal such thoughts to avoid intervention or hospitalization." In fact, just as Ross did, up to 78 percent of those who die by suicide explicitly deny suicidal thoughts in their last verbal communications before killing themselves. In hindsight, Jack believes his dad had every intention to end his life, even then. "When he made a decision, he made a decision," he told me. "So anything he's saying or doing is a decoy; it's going to be putting you off the scent."

To help clinicians determine which patients are most at risk of suicide, Nock and his associates have developed a clever predictive tool that works by revealing a person's camouflaged suicidal feelings. The tool is a variant of the Implicit Association Test (IAT), an established measure in the field of social psychology that identifies people's implicit—or hidden, and possibly even subconscious—mental associations about various topics. In Nock's version, participants sit down at a computer and see words associated either with "death" (such as *die, dead, deceased, lifeless, suicide*) or "life" (*alive, survive, live, thrive, breathing*) flash on the screen before them. Alongside these death- or life-related words are terms denoting either the attributes of "me" (*I, myself, my, mine, self*) or "not me" (*they, them, their, theirs, other*). The faster the participants push a button to link these concepts together, the stronger the mental association. Therefore, faster responding on the "death"/"me" blocks relative to the "life"/"me" blocks are taken to be suggestive of the person's suicidal feelings.

In 2010 Nock and his colleagues administered this five-minute computerized test to 157 adult patients who'd checked into a large psychiatric emergency department, most while they were still in the waiting room. Following up with these patients six months later, the predictive capability of this simple IAT procedure proved to be impressive. "The presence of an implicit association with death/suicide," the authors discovered, "was associated with an approximately six-fold increase in the odds of making a suicide attempt in the next six months." By contrast, the emergency room doctors who'd seen these patients got it completely wrong—despite their training, they failed to spot those most at risk of suicide.*

Administration of the IAT to gauge suicidal intent has yet to be adopted as standard clinical practice. Unfortunately, it wasn't done in Ross's case. Still, the psychiatrist had been concerned enough to recommend a term of daily home-care visits to Ross's house by a crisis counselor. One of the things weighing on everyone's mind was the sobering fact that Ross had made a suicide attempt twenty years earlier, when he was forty-five.

"I would have been in my early teens then," Jack told me of that earlier attempt. "He thought he was going to lose his job at the university. I remember him sitting at the dinner table, doing the maths, thinking out loud about his life insurance, and saying something like, 'If I die, you'll all be better off.' I guess he was thinking about some insurance windfall for his family. He'd gone into the garage and tried to do it with exhaust fumes, but my mum heard the car and got him out before anything happened. I think he went on holiday somewhere with my uncle, did a bit of fishing, came back, and it all worked out in the end."

"What do you mean?" I asked Jack. "How did the situation resolve itself?"

"They ended up appointing him senior lecturer."

"That's interesting," I point out. "I mean, at the time that he tried to kill himself back then, he must not have even considered the possibility that things would actually work out to his advantage. He must not have been able to see that as a potential outcome."

"Yeah," says Jack. "Apparently not."

This time, though, with his second marriage rapidly dissolving

*After treating the patient at the original intake session, the physician was asked the following question: "Based on your clinical judgment and all that you know of this patient, if untreated, what is the likelihood that this patient will make a suicide attempt in the next six months? (0–10, with 0 being no likelihood and 10 being very high likelihood.)"

and likely assault charges to boot, it would be a very different outcome. When the crisis counselor phoned to arrange a fifth home visit, she was surprised to hear Ross's estranged wife pick up.

Ross was dead, she'd said.

His sister had asked the authorities to do another welfare check, and when police arrived, they'd found Ross hanging in the garage. The police then called his wife, who'd just returned from abroad the evening before, and had her come over to identify the body.

Multiple versions of suicide notes, most of them meant for Ross's estranged wife, were carelessly left accessible on an old laptop at the scene—all of which were also sent to Jack on a memory stick.

"Some were clearly typed out in a rage," said Jack. "And some were well thought out. I think he meant her to see the more considered one, not the angry ones. But yeah, she probably saw all of them."

He showed me the notes. Even the more "watered-down" versions were pulsing with animus.

"It seems to me this upcoming court case was really getting to him," I told Jack, trying to look at the whole situation in context. "He must have felt very ashamed?"

"Oh yeah," Jack replied. "It's all through his emails, that he didn't think he had it in him. But he was so maddened by [his wife's deceit] that he lashed out. . . . He's got this hearing looming, ten days from when he died. And the final thing on that day, what really pushed him through the roof, was that he found out he was being appointed a senior barrister. And he's never been in trouble with the law before, you see. All of a sudden, he finds out he's going from a junior barrister to a senior barrister, and in his head, this was terrifying [because of the perceived seriousness of his offense]. He's going around in circles, saying, 'I wouldn't be in this mess if it wasn't for your lying' . . . that kind of thing. So he oscillates between seeing himself as the innocent party and the guilty party. At any rate, his second marriage was partly intended to spite the first, with my mum, but it was ultimately, in his mind, an epic fail. There was no hiding it. Word would get out."

From the very beginning, Jack has made it clear to me that he thinks his dad's decision was rational, logical. Yet poring over those suicide notes to his wife, I admit to Jack that "rational" isn't the first word that comes to my mind.

"What I mean by that," he explained to me, "is when people think suicide, they think mental illness straightaway, which is what annoyed me, because everyone was so quick to jump on that boat. There was a clear logic to what he did. He weighed up his options, he said, 'Here are my options: it's forgive and move on,' or forgive knowing that he's tried that before and it doesn't work in his head, that he wouldn't be able to live with it. So, he's weighed up Option A or Option B. It's 'here are the facts.' Suicide seemed to him a logical decision. Can't live with it, so may as well end it. I don't mean 'rational' as in a good decision. He did look at the future, and whatever way he looked at the future, it was blank."

Ending his life was rational *for Ross* is what Jack is trying to tell me.

"I am, I have concluded," writes Ross in one of his suicide notes, "an innocent, naïve, idealistic fool . . . I don't blame the world but my own inflexibility within it."

✷

Jack's analysis of his dad's decision making actually strikes at the heart of centuries of philosophical debate surrounding the ethics of suicide. And a heated debate it's been. Taking religion out of the equation, do we have the "right" to kill ourselves or even, in some cases, a moral duty? Or is suicide always, unequivocally, wrong? Is it always the sign of a mentally ill mind, or can it sometimes be characteristic of a rational and healthy one?

Today, even asking such questions evokes anxieties, and understandably so. But historically, the range of views on this gray matter has been vast and fierce. In ancient Greece, the Stoics treated suicide as the act par excellence of a cultivated thinker. Seneca famously wrote in his matter-of-factly titled "On the Proper Time to Slip the Cable," for instance, that "a wise man will live as long as he ought, not as long as he can."

> As soon as there are many events in his life that give him trouble and disturb his peace of mind, he sets himself free. . . . The best

> thing which eternal law ever ordained was that it allowed to us one entrance into life, but many exits. Must I await the cruelty either of disease or man, when I can depart through the midst of torture, and shake off my troubles? This is the one reason why we cannot complain of life: It keeps no one against his will. . . . Live, if you so desire; if not, you may return to the place whence you came. . . . If you would pierce your heart, a gaping wound is not necessary; a lancet will open the way to that great freedom, and tranquility can be purchased at the cost of a pin-prick.

And yet, more than fifteen hundred years after Pliny the Elder argued that suicide was a "supreme boon" and the one act that even the gods couldn't perform, the philosopher Immanuel Kant, who seems to have had a thorn in his paw on this particular topic, countered that it was, by contrast, the only thing that placed man lower than the animals. "If he disposes of himself," writes Kant, "he treats his value as that of a beast. . . . We are free to treat him as a beast, as a thing, and to use him for our sport as we do a horse or a dog, for he is no longer a human being; he has made a thing of himself, and, having himself discarded of his humanity, he cannot expect that others should respect humanity in him." It's almost as though, by making nonreligious moralistic arguments against suicide, scholars such as Kant are screaming at corpses in a futile attempt to validate their own existence. But they might as well protest the insouciance of stones; suicide is one of the few social acts for which, in the end, the individual does not have to face society.

Throughout the written ages, almost every influential thinker has at least ventured an opinion on the subject of suicide. From Plato being horrified by any man who would "slay his own best friend" to Nietzsche being consoled through many a dark night by the thought that suicide is always an option, most of the heavy hitters have chimed in. Ultimately, it's a matter of taste, really, which argument appeals to you most. Yet, as a psychologist, asking whether suicide is morally wrong or right feels almost irrelevant to me . . . and for the nihilist in me, it's a sort of vacuous query anyway. Is there any *inherent* reason to live a life assailed by troubles? Well, no.

Not really. As that straight-talking infidel of the British Renaissance David Hume put it in his essay on suicide, "the life of a man is of no greater importance to the universe than that of an oyster. . . . I am not obliged to do a small good to society, at the expense of a great harm to myself. Why then should I prolong a miserable existence, because of some frivolous advantage, which the public may, perhaps, receive from me?"

Still, I don't know about you, but I find myself in a difficult place when reading people like Seneca and Hume on suicide. Yes, they make sense. A little too much sense, if you know what I mean. Don't forget that when we're suicidal, our reasoning capabilities are seriously impaired in the ways we've seen, and as such, any level-headed attempt to crunch the numbers—expected future happiness quotients, quantified harm and benefits to loved ones, informed estimates of unexpected outcomes—can be subject to tragic miscalculations.

Clever as he was, Ross got the numbers wrong when he tried to kill himself at age forty-five. By his own admission, he had the best years of his life after that. And to my mind, at least, there's no reason to think that he didn't get it wrong when he actually did end his life at sixty-five.

It's not that there aren't rational suicides. There are; of course, there are. When twenty-year-old Ryan Lock, a young Brit who'd joined Kurdish freedom fighters in Syria, found himself surrounded by Islamic State militants in the northern city of Raqqa in December 2016, he turned his gun on himself. We all know what ISIS does to its hostages. But it would be a stretch to say that Lock was "suicidal" in the way that term is normally used.

I'm reminded of a passage from Valerius Maximus's *Memorable Doings and Sayings* in which the author describes what strikes me as a perfectly sane end-of-life custom among the Massilians (first-century inhabitants of what is now the South of France). Older individuals who were still in good health could preempt the ravages of age by appealing to the senate for permission to end their life. "The enquiry is conducted with firmness tempered by benevolence," wrote Valerius, "not suffering the subject to leave life rashly but providing swift means of death to one who rationally desires a way out." Typically, they'd be given poison compounded of hemlock, which was otherwise kept under close guard. Such a custom likely originated in Greece, the author surmises, and he recalls having once witnessed on the island of Kea the following extraordinary scene:

> A lady of the highest rank there but in extreme old age, after explaining to her fellow citizens why she ought to depart from life, determined to put an end to herself by poison. . . . Having passed her ninetieth year in the soundest health of mind and body, she lay on her bed, which was spread, as far as might be perceived, more elegantly than every day, and resting on her elbow she spoke: ". . . may the gods whom I am leaving rather than those to whom I am going repay you because you have not disdained to urge me to live nor yet to be witness of my death. As for me, I have always seen Fortune's smiling face. Rather than be forced through greed of living to see her frown, I am exchanging what remains of my breath for a happy end, leaving two daughters and a flock of seven grandchildren to survive me." Then, having urged her family to live in harmony, she distributed her estate among them, and having consigned her own observance and the domestic rites to her elder daughter, she took the cup in which the poison had been mixed in a firm grasp. After pouring libations to Mercury and invoking his divine power, that he conduct her on a calm journey to the happier part of the underworld, she eagerly drained the fatal potion. She indicated in words the parts of her body which numbness seized one by one, and when she told us that it was about to reach her vitals and heart, she summoned her daughters' hands to the last office, to close her eyes.

Recall Edwin Shneidman's maxim: "Never kill yourself while you are suicidal."*

*In his novel *The Comedians,* the author Graham Greene had a very different view: "However great a man's fear of life, suicide remains the courageous act, the clear-headed act of a mathematician. The suicide has judged by the laws of chance—so many odds against one that to live will be more miserable than to die. His sense of mathematics is greater than his sense of survival. But think how a sense of survival

"Well, fine," Jack interjected when I tried to argue this point in relation to his father's suicide. "But why do you assume that just because emotions are involved, one is irrational?"

That's the trouble with debating a philosopher.

Still, even Jack gets irked by the venerable members of his own pantheon. "These grand moralizers kind of get me incensed," he told me. "Because most of them are up in their ivory towers, and they've actually no bloody idea what it's like to be close [to suicide]. They're talking about stuff they don't know about. They just literally do not know. The objective, factual reality, what we are dealing with, involves actual states of mind and actual people and actual states of affairs. And they're all so different! You can't stretch a blanket over them with a fictional idea of what constitutes right and wrong."

*

The astonishing range of opinions regarding the ethics of suicide has implications beyond philosophy, too. Numerous studies have uncovered a correlation between a society's acceptability of suicide and its suicide rates. That is, even after adjusting for religious beliefs, those nations that tacitly endorse suicide as an individual right, or personal choice, have more suicides per annum than those in which the cultural attitude is more disapproving. This is why some suicide prevention experts caution that we should think twice before trying to eradicate the taboo of suicide entirely. I don't usually quote fiery early Puritan preachers favorably, but Increase Mather, responding to some clergy members wanting to reassess the church's unforgiving outlook on suicide, might have had a point when he wrote that "lest by being over-charitable to the dead, we become cruel to the living."

A few decades ago, the rabble-rousing anti-psychiatrist Thomas Szasz, best known for his book *The Myth of Mental Illness*, argued against suicide prevention—or at least, what he saw as the American tradition of forcibly interfering, often by involuntary hospitalization, with the death wishes of suicidal people. Even if it's a bad, impulsive

---

must clamor to be heard at the last moment, what excuses it must present of a totally unscientific nature."

decision, Szasz reasoned, suicide is a "fundamental right" and, aside from giving them counsel and helping them to see the error of their ways, we shouldn't impose our own iron will on someone who wants to end their life. No more than, say, we'd physically prevent someone we care about from making other bad decisions, like getting married to the wrong person. "The phrase 'suicide prevention,'" Szasz wrote in a 1986 article published in *American Psychologist*, "is itself a misleading slogan characteristic of our therapeutic age.... The term *prevention*, especially when coupled with *suicide*, implies coercion." Szasz suggested that a more effective approach would be a return to treating suicide as a moral, rather than a medical, question. Just because we have the right to kill ourselves, he argued, doesn't mean it's a good idea. "Where there is no freedom, there is no responsibility," Szasz explained.

> I object to our present policies of suicide prevention because they downgrade the responsibility of the individual (who is typically called a "patient" or "client" even if he or she rejects such a role) for the conduct of his or her own life and death.

In practice, however, morality isn't so straightforward. While Szasz was busy railing against suicide prevention efforts, the widow of a fellow psychiatrist who'd been seeing Szasz as a patient sued him over the death of her husband. Six months after Szasz had instructed the man to stop taking medication (lithium) for his bipolar disorder, the patient hanged himself, but that was only after he'd repeatedly struck himself in the head with a hammer and slashed his own neck.* The lawsuit against Szasz was settled out of court.

So how about you? Would you identify as a *moralist*, a *libertarian*, or a *relativist* when it comes to your feelings about the appropriate way to respond to a suicidal person? If you're a moralist, this means that you perceive suicide to be inherently wrong and that it should

*"It would seem certain that there are more cases like this in Szasz's files," Thomas Joiner—a vocal critic of Szasz's approach to suicide—speculates in his book *Mindlessness*.

be prevented at all costs. Libertarians, by contrast, are on the other side of the spectrum. For them, none of us is under any societal obligation to remain alive, and we certainly shouldn't be forced to live when, as autonomous free agents, we've carefully weighed our options and have concluded that death is the lesser harm. So-called zero suicide initiatives, in the view of many libertarians, are overly "paternalistic," another one of those morally loaded terms. Finally, the relativist says, in essence, "Well, it depends."

Previously, I asked Roy Baumeister if he thought his six-stage model of suicidal thinking, which we saw in chapter 4, could be a valuable tool for those working in the field of suicide prevention.

"Well," Roy hedged, "if we say that suicide is an outcome that should be prevented at all costs, that there's no justification, people should never kill themselves, then yes, knowing the process is useful. I was never quite persuaded by that. Maybe it's my libertarian streak." He continued:

> I remember going to a conference once. I was invited to give a talk on suicide and, going in, there were all these signs and posters about having a vigil and a meeting about how to prevent suicide. My wife was with me and I said, "Oh, okay, I guess they're against it." In other words, I'm trying to understand it but I never took the position that I'm unequivocally against suicide or trying to prevent all people from killing themselves. I don't know what the moral duty is. If you talked a person out of suicide and the person went on to be miserable for many years, that would sort of be your fault, you inflicted all that suffering on that person. So, I don't presume to know what's right for other people.

*

I once heard a disturbing story about a sleep-deprived cardiologist who, after a stressful day at work, became unhinged and bludgeoned his own pet dog in the backyard with a croquet mallet when it wouldn't stop barking. Then, realizing the gravity of what he'd done, the man went into his house and shot himself in the head.

I'm not proud of it, but my initial thought on hearing this account was not one of compassion for the human life lost. Instead, because dogs are my emotional weakness, I saw the man's suicide as a sign of a just world. "What a bastard," I thought. "Good riddance."

But that's the thing about people who kill themselves over some heinous transgression. It's only after they've done so that we're able to realize that, feeling as they so clearly did about their misdeeds, these were precisely the types of redeemable people who shouldn't have ended their lives this way. After all, it's those who aren't particularly bothered by their harming of others—or perhaps they simply don't care that the world knows what they've done—who are far more likely to harm others again.

It's a pity that the scales of justice tilt so heavily in our minds in favor of a person's worst-ever deed; the accumulated weight of their whole complex life story, including all the good they've ever done, becomes, in comparison, morally irrelevant. It was brutal and unspeakably cruel what he did to that dog, yes, but how many other lives had that cardiologist saved, or might have saved to come? My mind, admittedly, has trouble with such math.

Likewise, Jack's father did a despicable deed by striking his wife in the heat of the moment. Domestic violence is shameful, and Ross's behavior was arguably unforgivable. Victims of intimate partner assault are at a significant risk of suicide themselves. And yet, isn't reducing the entirety of Ross's identity to that split-second decision a shallow measure of the man? "I just blew a gasket," Ross wrote in his farewell letter to his son. "I did a bad thing frightening her and I am sorry."

Shortly before his suicide, Ross had designed a groundbreaking electromagnetic device that could reliably detect car bombs from a safe distance, an innovation that promised to save countless law-enforcement and civilian lives in war-torn regions across the globe. It's a project that would come to nothing without his direction and expertise. As Jack and others continue to grieve Ross's loss, he's remembered as an ingenious but imperfect man whose life, and potential, were worth cumulatively far more than his death.

"Dad was a huge part of my life," Jack said. "A sort of golden aspiration. When he went, I felt like I had no point in living anymore, either."

In the end, I think, we can philosophize about the ethics of suicide—and if there are ever any social situations that would make it a rational or logical act—as long as there are flawed human beings who find themselves suddenly, unexpectedly, facing a terrifying future in which their reputation and identity has been permanently stained by an act they regret. But love isn't philosophy. And when we truly love someone, we would do anything to persuade that person that suicide is not, in fact, the only option. The truth is, we'd hold that fallen person's hand through Hell on earth.

We are their other option.

✷

I've always had a thing about cemeteries. As a kid, my father would sometimes give into my begging and take me to a small derelict graveyard at the top of a hill just behind our neighbor's home. I couldn't have been more than six or seven at the time. We were living in a D.C. commuter suburb of Virginia then. The cemetery was a tiny windswept enclave where the residents had, so far, managed to fend off the urban sprawl swallowing up their once-remote town, filling their plentiful pastures with cheaply manufactured homes, and bulldozing their forests to make room for soccer fields and throughways into the capital.

Sure, the cemetery was overgrown and it was hard not to trip over the fast-food wrappers and empty beer cans left by the teens who liked to hang out there and smoke pot, but to me, this was a magical place. It was a fellowship of bones and secrets and sorrows. There I'd stand, squeezing my dad's hand while gazing with a kind of innate wonder at the drab white stones that reminded me of broken teeth.

The stones of stories lived.

One of these visits stands out to me now.

"Look," said my father, "this little boy wasn't much older than you when he died."

"How did he die?" I asked.

My father just shrugged.

Still, this unexpected truth, that death seized not only the frail and old, but even those everlasting ones—children like *me*—must have sat uncomfortably in my subconscious. Because it was around this same time, too, that I had the dream.

Many of us have that one special dream from our childhood . . . that one terrifying but mysterious journey into the numinous, that nightmare whose images and the powerful emotions they evoked we can still recall as clearly today as the moment we awoke, startled, in its raw wake, our hearts thumping in our chests and our thoughts alive with the echoes of its horrible wonder. Such dreamscapes are mostly ineffable things, since they are felt by the child at an almost cellular level, and any attempt to communicate them to others, to carve out their power in words, are frustratingly ineffective. That said, I'll do my best to share mine here, because the dream seems to have forecasted the contours of my life story as a psychologist, how I view the problems of being human, and my approach to suicide.

In this dream of mine, I was the sole audience member seated center stage in a grand old theater of gilt and rose. Although my recollections are undoubtedly colored by my adult imaginings, as I waited there, restless with anticipation, I remember hearing distant voices and the tinker of musical instruments, an invisible orchestra preparing for the opening scene. The lights dimmed, the orchestra reared up with a dramatic overture, and the curtains parted to reveal a very familiar-looking stage setting. There on theatrical display was a reconstruction of my childhood bedroom, but one in which the artificiality of the scene was clearly meant to be noticeable, in that aspects of hyperrealism were deliberately offset by cardboard features. There in that cardboard bed was the character of me, sleeping soundly and dreaming, well, the very dream I was dreaming.

As I observed this oddly emotive scene from my vantage point in the audience, the atmosphere began to change . . . the interstitial music darkened in tone, it felt as though there were a precipitous drop in altitude, and I became aware of something unpleasant about to happen. That's when I saw it. Through the stage bedroom window, as the oblivious me slept undisturbed, a ghost hovered just outside. Backlit by the glow of a cardboard moon, it peered in at the me in bed, floating in space, rapping ever so delicately against the glass. It wasn't just any ghost, either. It was the ghost of me. That is to say, me too. Or rather, me three.

I awoke screaming. My parents rushed in. But it was an unassuageable fright.

Our minds are magpies that gather up random bits and pieces of our day-to-day lives, weaving together the emotional litter, subliminal percepts, and cognitively processed scraps to create bedazzling nocturnal stories and images. When pulled apart, my dream was certainly no more than that, perhaps made up of those cemetery visits, which also happened to coincide with my supporting role as a Hanukkah dreidel in a first-grade winter play. But even if I forgo any supernatural explanations for the dream, this strangely visual portrayal of my splintered ego has nonetheless stuck with me all these years. I've summoned it as a metaphor on occasion, as it helps me to see things in a different light. And sometimes, when we're contemplating suicide, a shift in perspective is exactly what we need.

Consider the angle of the dream. Yes, there are three versions of me—the one in the audience, the one onstage, and the one beyond—but I could see things only from my point of view. That is, my perspective in the dream was that of the onlooker, the solitary audience member watching from afar. With only one set of eyes, I could not simultaneously be the character on the stage *and* the observer watching the character on the stage *and* my ghostly self. Sometimes, when it comes to dealing with the most difficult moments of our lives, I think this is precisely where we should try to position ourselves: in the audience, not on the stage, or beyond. We feel the emotions of the actors as if they are ours—that's the object of good theater, after all, to stir up intense feelings, so much so that the shock may even startle us awake in our seats. This ongoing drama of our lives provides us with meaning and we're happily deluded by it. But the spectator is also aware of his separateness. He has perspective. He is, in a sense, untouchable from the chaos of the act he observes.

Usually, this stage-seating effect happens automatically as a natural defense mechanism. When we're in a suicidal state, however, we temporarily lose that reflective, lifesaving outer eye of the interested onlooker who is simply there to see where the story goes. Instead, we become our characters, trapped on the stage, overwhelmed by affect, and myopic in our vision. As such, we are vulnerable to the chaos that consumes them; we become acutely aware of those cardboard edges, the threadbare stitching of the costumes being worn, the whole enveloping artifice of the social roles that we play. We forget that the frightening presence loitering outside that promises a life beyond the already impossible, fantastical one we've been given is in fact part of the set as well, a specter illuminated by a cardboard moon.

What we sometimes must remind ourselves is that the scene will change. It always does. And although we can't always consciously intervene in the process and insert ourselves back into the audience to watch our own life story playing out, we will, sooner or later, find ourselves in that safe space again. This is an important point for those of you who, like me, are simply *incapable* of finding succor in religion or any other belief system in which human suffering is conceived as meaningful . . . that is to say, with some intelligent agency behind it.

"I don't know how you can go on thinking life is meaningless," someone said to me recently. "Seriously, what's the point of even getting up in the morning if there's no purpose to it all?" But I can no sooner choose to believe than others can choose not to. If I had any say over the matter, I'd hold a very different view of reality. I'd choose to believe that our fates were indeed meant to be, that everything happens for a reason. Besides, being an atheist is like being surrounded by people who are in the middle of a role-playing fantasy game and you can't get them to just be serious for a minute—and, really, who wants to be a buzzkill?

The thing is, it doesn't matter what we believe; when we're emotionally healthy, we can't help but live our lives *as if* they're meaningful. It just . . . happens. We may know better, but we're still beautifully deluded. That's the power of the human brain; it's what helps us sober rationalists drag our absurd carcasses out of bed every day and keeps us from jumping off the nearest bridge at the slightest inconvenience. When it comes to suicide, the existence or nonexistence of God is far less consequential than the tumult of our social lives.

In the end, there are no satisfying answers. That is the truth about suicide. Still, we need to find a way to live with the inherent ambiguity of the subject matter. And that—living comfortably with uncertainty—is something of a lost art. For some, suicide is a tragedy; for others, it may offer salvation. But to pretend it's anything other than morally complex will do none of us any favors. So, let us scratch our heads gracefully. Perhaps that's the best and most honest thing we can do when it comes to the philosophical quandaries of suicide.

If we want to actually save lives, including our own, that's another thing entirely. There is much we can do. On the more practical side, be aware of environmental triggers—literally. In some parts of the U.S., where the chances of finding a gun in the home (the place where, incidentally, most suicides occur) are about as good as finding a carton of milk in the fridge, there's no better predictor of suicide than simply having access to a firearm. In a yearlong study of new California gun purchasers, suicide was the leading cause of death, accounting for 25 percent of that group's fatalities. Although women aren't as likely to go out and buy a handgun, when they do, they're more likely to turn it on themselves; of the female subset of those deceased California gun owners, over half were suicide victims.

Controlling access to the means of suicide has demonstrable effects on suicide prevention. In one Inuit community in the Canadian Arctic about a decade ago, there were so many hangings among teenage boys that the local housing authority decided to remove the closet rods from every home. "Hanging in most cases took place at home during the night when the family was asleep," explained the psychologist Michael Kral, who studied this group of indigenous people. "[It was] from the closet bar with the clothes pushed to the right and the noose tied on the left side . . . with the victim facing the wall." Suicide-proofing the home environment this way might sound too simple, some would say impractical, to be very effective. But it worked. According to Kral, the town went from having the highest rate of suicide in the region to *zero* suicides four years running. Adding catalytic converters to the standard vehicle exhaust system has made the old closed garage-door suicide method almost obsolete. In European countries where suicide by drowning was a major problem, mandatory swimming lessons for young children in the 1980s has correlated with fewer such adult deaths in recent years. Bridge barriers erected at suicide jumping hot spots in New Zealand and Australia have also proven effective deterrents.

The point is that when you make it a pain for people to die by their first-choice method, they don't necessarily just find another way to go about it. This is especially true for impulsive, unplanned suicides, the type most often found among children and adolescents. "Many young suicide attempters report that they spent only minutes between the decision and the actual attempt," explains the psychiatrist Urs Hepp. "As impulsive suicide attempters use more violent methods, such as firearms, hanging and jumping, this [shows] the importance of restricting access to highly lethal methods."

Another important lesson that researchers in this area have learned is that what makes someone "successful" isn't success, but how they cope with failure. One of the most valuable things we can do for children is to help them navigate emotionally through their various disappointments, not prevent their having them. Shielding children from all misfortune does them not only a disservice, but potentially places them at risk for maladjustment when the inevitable major crisis arises in their adult lives. Perfectionism can become deadly. So, if you've got kids or plan to have them, let them fail from time to time, and teach them that it's part and parcel of being human. Failure is a gift that may save their lives someday.

Finally, as I've tried to argue throughout this book, I believe we're also better equipped to face our own periodic suicidal feelings simply by remaining ever aware that we are social animals—cerebral beasts whose mental machinery has evolved to be so finely attuned to the stuff of other people's thoughts that sometimes our very existence hangs in the balance of what we think others think of us. This quirk of our psychology, of being the natural psychologists of the animal world, is the ultimate burden of being human—the thorn in the crown of our consciousness. Although there's no way to "turn off" this basic social cognitive function so that our emotions are no longer a product of the opinions and judgments of other people, we can at least seek out like-minded allies who value and appreciate us. Sometimes, in fact, one is all we need: a single fellow human being who acknowledges our social suffering. In short, a friend who ventures to love us despite us.

It works the other way too, of course. Consider, in closing, a cautionary tale for strangers. (Never underestimate the effect of altruistic strangers; a discrete kindness from someone we don't know has mysterious healing powers.) You may have heard this story before as some apocryphal offering, but in fact it's true, as verified by a journalist in the *New Yorker* as happening back in the seventies. After a lonely Bay Area man killed himself by jumping off the Golden Gate Bridge, his psychiatrist and the county's assistant medical examiner entered the dead man's apartment. A suicide note had been left on a bureau. "I'm going to walk to the bridge," it read.

"If one person smiles to me on the way, I will not jump."

acknowledgments

This book began in angst. The irony is that my idea to write a book about suicide first came about while I was having suicidal feelings, which were there in no small part because I'd run out of ideas for my next book. More precisely, I'd run out of money. Anyway, what's that old saying? "Luck is what happens when preparation meets opportunity." That's pretty much how I got here. Luck, in this case, meant getting the chance to write this book instead of killing myself. I was prepared and had plenty of opportunities to do things the other way around too—because don't forget, these two factors are catalysts not just for luck, but also suicide. But I'm glad it worked out this way. What more can I say? Life is fickle.

I don't mean to sound grim. It's just the topic. The truth is, I greatly enjoyed working on this book. It was—to use that overused term—cathartic. Also, sometimes, when you force yourself to think about unpleasant things, day and night, and in an academic sense, a strange beauty begins to reveal itself in their intricacies. That's not to romanticize something so terrible as suicide; for those left in its wake, it is devastating. Rather, I mean only to say that there is order in the chaos, and when you begin to understand the intricate clockwork of suicide—when you know what makes a human being tick in this fashion—you'll never quite look at it the same way again. That's a good thing; a shift in perspective can be life altering.

Many people helped get this book out of my head and into your hands. At the top of the list is Peter Tallack, my agent at the Science

Factory, who was patient enough to see the proposal through to fruition. Alongside him was Tisse Takagi, who helped shape the scope of this work. The hardworking group at the University of Chicago Press has been enormously supportive of this project from its inception. I'm especially grateful to Garrett Kiely for seeing the potential value of this undertaking, Christie Henry for giving it wings with early editorial work, and Priya Nelson, who bravely took over as editor and was instrumental in helping me to develop the manuscript. Levi Stahl did the marketing with gusto, Erin DeWitt was a marvel with copyediting, and Dylan Montanari did the legwork with getting the final bits and pieces in order. I'm also indebted to two anonymous reviewers who offered detailed comments on an earlier draft. And in the UK, Doug Young and his fantastic team at Transworld were similarly with me from the start; I am very fortunate to have also had their unwavering support. This work was also made possible through a generous Prestigious Writing Grant from my employer, the University of Otago.

For those researchers, advocates, and other professionals who granted me interviews or responded to my emails about their important work, thank you. To name a key handful: Denys DeCatanzaro, Roy Baumeister, Carrie Jurney, David Chalmers, and Joe Nidd. Thanks also to the religious leaders who participated in my roundtable discussion about suicide. Bonnie Scarth, my tireless PhD student, provided encouragement and healing conversation during the writing of this book, and those who volunteered their private experiences and insight through Bonnie's interview sessions for her doctoral thesis added greatly to this book.

Numerous friends and colleagues read and gave me feedback on various portions from early incarnations of this book. My sincere thanks goes out to Jenny Rock, Todd Shackelford, Lloyd Spencer Davis, Sue Harvey, Nancy Longnecker, Fabien Medvecky, Ross Johnston, Mary Roach, Jamin Halberstadt, Harvey Whitehouse, Emma Curtin, and Christopher Ryan. Also Nikki Saadat, who provided early assistance with the literature review. My students at the Centre for Science Communication—namely, Michelle Walsh, Emma Harcourt, Elise Provis, Vibhuti Patel, and Conor Feehly—

helped me to stay the course with their inquisitiveness and conversations along the way. With perpetual gratitude, I thank my partner, Juan Quiles. It's not easy living with someone who is writing a book about suicide for eighteen months straight. It's out of my system now, I promise.

Most of all, I am indebted to those who had the courage and candor to share their personal stories about the loss of their loved ones to suicide. Thank you for your faith and trust in me. I will forever be in awe of your strength and clear wisdom in the face of such immeasurable grief.

resources

United States

Suicide Prevention Lifeline (1-800-273-8255; https://suicidepreventionlifeline .org/) is a 24-hour, toll-free, confidential suicide prevention hotline available to anyone in suicidal crisis or emotional distress.

The *American Foundation for Suicide Prevention* (https://afsp.org/) is a nonprofit organization exclusively dedicated to understanding and preventing suicide through research, education, and advocacy, and to reaching out to people with mental disorders and those impacted by suicide.

Crisis Text Line (https://www.crisistextline.org/) is the only 24/7 nationwide crisis-intervention text-message hotline.

The Trevor Project (http://www.thetrevorproject.org/) is a nationwide organization that provides crisis intervention and suicide prevention to lesbian, gay, bisexual, transgender, and questioning youth.

Australia

Lifeline (https://www.lifeline.org.au/) is a 24-hour nationwide service that provides access to crisis support, suicide prevention, and mental health support services. It can be reached at 13-11-14. They also offer an online chat service.

Beyond Blue (https://www.beyondblue.org.au) provides nationwide information and support regarding anxiety, depression, and suicide. It has a helpline that can be reached by calling 1300-22-4636. The helpline is available 24 hours a day, 7 days a week. In addition, the organization provides an online chat.

Canada

The *Canadian Association for Suicide Prevention* (http://suicideprevention.ca/) provides information and resources to reduce the suicide rate and minimize the harmful consequences of suicidal behavior. CASP maintains an up-to-date list of Distress Lines across Canada by province and territory.

Kids Help Phone (https://kidshelpphone.ca) is a 24-hour, toll-free, confidential crisis line (1-800-668-6868) and online counseling service available to Canadians under the age of twenty.

Suicide Action Montréal (http://suicideactionmontreal.org/) offers support services, crisis workers, and monitoring for people who are at risk of committing suicide, for their friends and family, and for people affected by suicide.

China

Beijing Suicide Research and Prevention Center Hotline (http://www.crisis.org.cn) is available 24/7 and can be reached at 800-810-1117 or 010-82951332.

Lifeline Shanghai (http://www.lifeline-shanghai.com; 021-6279-8990) is a volunteer-based, English-speaking organization for the international community. They are open from 10 a.m. to 10 p.m. every day. They offer a live chat service.

Ireland

Samaritans (http://www.samaritans.org/) is a registered charity aimed at providing emotional support to anyone in distress or at risk of suicide throughout Ireland.

Japan

TELL Lifeline (http://telljp.com/lifeline/) provides a free, confidential English-language Lifeline service, plus clinical mental health services, for the international community in Japan. They can be reached at 03-5774-0992, 9 a.m.–11 p.m., every day.

Befrienders Worldwide, Osaka Suicide Centre (http://www.spc-osaka.org).

Mexico

SAPTEL (http://www.saptel.org.mx/index.html) is an independent care provider subsidized by the Mexican Red Cross. It is totally free, and they are available 24 hours a day: (55) 5259-8121. Crisis dialogue or treatment for anything related to mental health crisis.

New Zealand

Youthline (https://www.youthline.co.nz/) is a free service that provides a safe and accepting environment to give you the time and space that you need to talk if you want to. You do not have to be in a crisis situation to ring. Call: 0800-376-633 for 24/7 support. Free text: 234 between 8 a.m. and midnight. Email: talk@youthline.co.nz.

Lifeline Aotearoa (http://www.lifeline.org.nz) provides free 24-hour counseling and phone help lines (0800-543-354). It provides support, information, and resources to people at risk of suicide, family and friends affected by suicide, and people supporting someone with suicidal thoughts and/or suicidal behaviors.

United Kingdom

PAPYRUS Prevention of Youth Suicide (http://www.papyrus-uk.org) is a national charity for prevention of youth suicide. It operates a helpline (0800-068-41-41) providing short-term support and advice for anyone concerned about themselves or a young person they know.

Samaritans (https://www.samaritans.org/) is a charity aimed at providing emotional support to anyone in distress or at risk of suicide throughout the United Kingdom.

For a list of other suicide help organizations around the globe, see https://en.wikipedia.org/wiki/List_of_suicide_crisis_lines.

notes

EPIGRAPH

vii "And so far forth": Robert Burton, *Anatomy of Melancholy* (1621; New York: New York Review Books, 2001), 431.

CHAPTER ONE: THE CALL TO OBLIVION

1 "Just as life had": Virginia Woolf, *The Death of the Moth and Other Essays* (1942; New York: Harcourt Brace & Company, 1970), 8.

2 "dizzying in their variety": Thomas E. Joiner et al., "Suicide as a Derangement of the Self-Sacrificial Aspect of Eusociality," *Psychological Review* 123, no. 3 (2016): 240.

2 "not suicide [but] self-deliverance": Derek Humphrey, *Final Exit: The Practicalities of Self-Deliverance and Assisted Suicide for the Dying* (1991; New York: Delta, 2010), 90–91.

3 The vast majority of those: Kay Redfield Jamison, *Night Falls Fast: Understanding Suicide* (New York: Vintage, 2000).

3 According to one estimate: John M. Bostwick and V. Shane Pankratz, "Affective Disorders and Suicide Risk: A Reexamination," *American Journal of Psychiatry* 157, no. 12 (2000): 1925–32.

3 "psychache," he called it: Edwin S. Shneidman, *The Suicidal Mind* (Oxford: Oxford University Press, 1996).

7 One of the cruelest tricks: Roy F. Baumeister, "Suicide as Escape from Self," *Psychological Review* 97, no. 1 (1990): 90–113.

7 our well-being is hugely dependent: Mark R. Leary and Roy F. Baumeister, "The Nature and Function of Self-Esteem: Sociometer Theory," *Advances in Experimental Social Psychology* 32 (2000): 1–62.

7 Social psychologist Roy Baumeister: Roy F. Baumeister, "Suicide as Escape from Self," *Psychological Review* 97, no. 1 (1990): 90–113.

7 "A reverse of fortune": Madame de Staël, *Reflections on Suicide* (1813), in *The Constitution of Man, Considered in Relation to Eternal Objects*, Alexandrian Edition, ed. George Combe (Columbus, OH: J & H Miller, n.d.), 100.

8 "I felt, I hesitate to admit": Albert Camus, *The Fall*, trans. Justin O'Brien (1956; New York: Vintage, 1991), 29.

13 "We had all forgotten that": Fernando Pessoa, *The Book of Disquiet*, trans. Margaret Jull Costa (1982; London: Serpent's Tail, 2011), 30.

13 using a projective test: Edwin S. Shneidman, "The Make-A-Picture-Story (MAPS) Projective Personality Test: A Preliminary Report," *Journal of Consulting Psychology* 11, no. 6 (1947): 315–25.

15 statistically we're far more likely: Kristen L. Syme et al., "Testing the Bargaining vs. Inclusive Fitness Models of Suicidal Behavior against the Ethnographic Record," *Evolution & Human Behavior* 37, no. 3 (2016): 179–92.

16 "always being free to": Lewis Cohen, "How Sigmund Freud Wanted to Die," *The Atlantic*, September 23, 2014, https://www.theatlantic.com/health/archive/2014/09/how-sigmund-freud-wanted-to-die/380322/.

16 "Never kill yourself while": Edwin S. Shneidman, *The Suicidal Mind* (Oxford: Oxford University Press, 1996), 166.

16 rarely last longer than: Ibid.

17 "one truly serious philosophical": Albert Camus, *The Myth of Sisyphus*, trans. Justin O'Brien (1942; New York: Vintage, 1991), 3.

CHAPTER TWO: UNLIKE THE SCORPION GIRT BY FIRE

18 "For all cats have": Angela Carter, *The Bloody Chamber: And Other Stories*, 75th Anniversary Edition (1979; New York: Penguin Books, 2015), 85.

20 "act with a fatal outcome": American Psychiatric Association, *Diagnostic and Statistical Manual of Mental Disorders*, 5th ed. (*DSM*-5) (Washington, DC: American Psychiatric Publishing, 2013).

21 "higher psychological processes": C. Lloyd Morgan, *An Introduction to Comparative Animal Psychology*, 2nd ed. (London: W. Scott, 1903), 59.

22 as the science historians: Edmund Ramsden and Duncan Wilson, "The Suicidal Animal: Science and the Nature of Self-Destruction," *Past & Present* 224, no. 1 (2014): 201–42.

22 "is like the Scorpion girt by fire": Lord Byron, *The Giaour: A Fragment of a Turkish Tale* (London, 1813).

23 "Suicide of Scorpions": C. Lloyd Morgan, "Suicide of Scorpions," *Nature* 27 (1883): 530.

23 scorpions are immune: San Diego Zoo, "Scorpions," n.d., http://animals.sandiegozoo.org/animals/scorpion.

24 "operations of my own individual": George J. Romanes, *Animal Intelligence* (London: Keagan Paul, 1882), 1–2.

24 "no fundamental difference": Charles R. Darwin, *The Descent of Man, and Selection in Relation to Sex: The Concise Edition,* selections and commentary by Carl Zimmer (1871; New York: Plume, 2007), 110.

24 packaged in the frontal cortex: Todd M. Preuss, "Who's Afraid of *Homo sapiens*?" *Journal of Biomedical Discovery and Collaboration* 1, no. 17 (2006): 17.

25 "human cognitive specializations": Daniel J. Povinelli and Todd M. Preuss, "Theory of Mind: Evolutionary History of a Cognitive Specialization," *Trends in Neurosciences* 18, no. 9 (1995): 418–24.

25 being a bat: Thomas Nagel, "What Is It Like to Be a Bat?" *Philosophical Review* 83, no. 4 (1974): 435–50.

25 what makes us unique: Michael Tomasello, *A Natural History of Human Thinking* (Cambridge, MA: Harvard University Press, 2014).

26 "natural psychologists": Nicholas K. Humphrey, *The Inner Eye: Social Intelligence in Evolution* (Oxford: Oxford University Press, 1986).

26 crocodile-infested waters: " 'You Can't Legislate against Human Stupidity': Crocodile Attack Victim Blamed for Her Own Death," *National Post,* May 30, 2016, http://nationalpost.com/news/world/you-cant-legislate-against-human-stupidity-crocodile-attack-victim-blamed-for-her-own-death.

26 "You can't legislate": Ibid.

27 "All I could think about": Diana Sands and Mark Tennant, "Transformative Learning in the Context of Suicide Bereavement," *Adult Education Quarterly* 60, no. 2 (2010): 111.

29 "The person who commits suicide": Edwin S. Shneidman, *On the Nature of Suicide* (San Francisco: Jossey-Bass, 1969), 22.

29 "She had carefully disguised": Al Alvarez, *The Savage God: A Study of Suicide* (1971; London: Bloomsbury, 2002), 49.

30 autoeroticism gone bad: Riazul Imami and Miftah Kemal, "Vacuum Cleaner Use in Autoerotic Death," *American Journal of Forensic Medical Pathology* 9, no. 3 (1988): 246–48.

30 "complex sleep-walking": Mark W. Mahowald et al., "Parasomnia Pseudo-Suicide," *Journal of Forensic Sciences* 48, no. 5 (2003): 1158–62.

31 "yellow-haired bulldog": "Dog a Suicide? Partly Blind Animal Finds Death in Hippo's Tank," *Washington Post,* September 2, 1913, 4.

31 "thought it had lost its master": "War Talk Kills This Dog: Animal Commits Suicide at the Thought of Separation from Its Master," *Washington Post,* September 5, 1914, 6.

31 a cat named Topsy: "Bereaved Cat Commits Suicide," *Washington Post,* March 21, 1911, 6.

32 lovesick pet monkey: "Unhappy Hooligan: Pet Monkey Committed Suicide in Chicago," *Boston Daily Globe*, June 29, 1907, 6.

32 wily capuchin named Bok: "Monkey Commits Suicide: Reprimanded by Slaps on the Back He Kills Himself," *Washington Post*, February 24, 1906, 6.

32 A wan cat: "Cat Commits Suicide," *Washington Post*, November 25, 1906, RA8.

32 belabored draft horse: "The Suicide of a Horse: Exhausted by Cruelty, He Leaped over a Precipice," *Washington Post*, September 18, 1898, 14.

33 substrates of physical and emotional: Patrick Bateson, "Assessment of Pain in Animals," *Animal Behaviour* 42, no. 5 (1991): 827–39.

34 same calming hormone: Gerald Gimpl and Falk Fahrenholz, "The Oxytocin Receptor System: Structure, Function, and Regulation," *Physiological Reviews* 81, no. 2 (2001): 629–83.

34 "When people reject": Edmund Ramsden and Duncan Wilson, "The Suicidal Animal: Science and the Nature of Self-Destruction," *Past & Present* 224, no. 1 (2014): 240–41.

36 true psychopaths rarely: Hervey M. Cleckley, *The Mask of Sanity: An Attempt to Clarify Some Issues about the So-Called Psychopathic Personality*, 2nd ed. (1941; Eastford, CT: Martino Fine Books, 2015).

36 Subsequent research suggests: Edelyn Verona, Christopher J. Patrick, and Thomas E. Joiner, "Psychopathy, Antisocial Personality, and Suicide Risk," *Journal of Abnormal Psychology* 110, no. 3 (2001): 462–70.

36 "The almost total lack of": Michael J. Garvey and Frank Spoden, "Suicide Attempts in Antisocial Personality Disorder," *Comprehensive Psychiatry* 21, no. 2 (1980): 148.

37 "VENS, it seems": Martin Brüne et al., "Neuroanatomical Correlates of Suicide in Psychosis: The Possible Role of von Economo Neurons," *PLOS ONE* 6, no. 6 (2011): 4.

38 "The ability to reflect upon": Ibid., 4.

38 "Shame was the most terrible": Andrew Rankin, *Seppuku: A History of Samurai Suicide* (Tokyo: Kodansha International, 2012), 159–60.

39 "Lo, my name is abhorred": Chris Thomas, "First Suicide Note?" *British Medical Journal* 281, no. 6235 (1980): 284–85.

39 In his play: Jean-Paul Sartre, *No Exit and Three Other Plays* (1944; New York: Vintage, 1989).

40 "Open the door!": Ibid., 41.

40 "We essentially live": Philippe Rochat, "Commentary: Mutual Recognition as a Foundation of Sociality and Social Comfort," in *Social Cognition: Development, Neuroscience, and Autism*, ed. Tricia Striano and Vincent Reid (Malden, MA: Wiley Blackwell, 2009), 306.

41 "it's too early": Marc Bekoff, *Why Dogs Hump and Bees Get Depressed: The*

Fascinating Science of Animal Intelligence, Emotions, Friendship, and Conservation (Novato, CA: New World Library, 2013), 168.

41 “Naturalists have not identified”: Antoni Preti, “Suicide among Animals: A Review of Evidence,” *Psychological Reports* 101 (2007): 831.

42 myth of the leaping lemmings: David Mikkelson, “Did Disney Fake Lemming Deaths for the Nature Documentary ‘White Wilderness’?” *Snopes*, May 23, 2017, http://www.snopes.com/disney/films/lemmings.asp.

42 “get down to a dog’s level”: “Why Have So Many Dogs Leapt to Their Deaths from Overtoun Bridge?” *Daily Mail*, October 17, 2006, http://www.dailymail.co.uk/news/article-411038/Why-dogs-leapt-deaths-Overtoun-Bridge.html.

43 “billowy white pile”: “Sheep, or Lemmings?” *Wired*, July 7, 2005, https://www.wired.com/2005/07/sheep-or-lemmings/.

43 startled by a pack of wolves: John Lichfield, “This Europe: Shepherds Despair as Wolf Packs Drive Sheep to Suicide,” *Independent*, July 24, 2002, http://www.independent.co.uk/news/world/europe/this-europe-shepherds-despair-as-wolf-packs-drive-sheep-to-suicide-5361128.html.

43 Whale beachings have also: Bala Sundaram et al., “Acoustical Dead Zones and the Spatial Aggregation of Whale Strandings,” *Journal of Theoretical Biology* 238, no. 4 (2006): 764–70.

44 most commonly cited example: Joanne P. Webster and Glenn A. McConkey, “*Toxoplasma gondii*-Altered Behaviour: Clues as to Mechanism of Action,” *Folia Parasitologica* 57, no. 2 (2010): 95–104.

44 My own favorite: D. G. Biron et al., “ ‘Suicide’ of Crickets Harbouring Hairworms: A Proteomics Investigation,” *Insect Molecular Biology* 15, no. 6 (2006): 731–42.

44 human beings infected with: Midori Tanaka and Dennis K. Kinney, “An Evolutionary Hypothesis of Suicide: Why It Could Be Biologically Adaptive and Is So Prevalent in Certain Occupations,” *Psychological Reports* 108, no. 3 (2011): 977–92.

46 “Nonhuman animals also have”: Laurel Braitman, *Animal Madness: Inside Their Minds* (New York: Simon & Schuster, 2014), 168.

46 “the sadness into which”: Émile Durkheim, *Suicide: A Study in Sociology*, trans. John A. Spaulding (1897; New York: Free Press, 1997), 45.

CHAPTER THREE: BETTING ODDS

48 “Sometimes it’s better”: Terry Pratchett, *Men at Arms* (1993; New York: Harper, 2013), 282.

51 fanciful theories of human nature: Desmond Morris, *The Naked Ape: A Zoologist’s Study of the Human Animal* (1967; New York: Random House, 2010).

51 "will never produce in a being": Charles Darwin, *On the Origins of Species by Means of Natural Selection* (1859; Terceira Açores, Portugal: Erres e Esses, Lda., 2009), 150.

52 a young British polymath: William D. Hamilton, "The Genetical Evolution of Social Behaviour. II," *Journal of Theoretical Biology* 7, no. 1 (1964): 17–52.

53 such a perspective materialized: Edward O. Wilson, *Sociobiology: The New Synthesis*, 25th Anniversary Edition (1975; Cambridge, MA: Harvard University Press, 2000).

53 a sort of illusory puppetry: Richard Dawkins, *The Selfish Gene*, 40th Anniversary Edition (1976; Oxford: Oxford University Press, 2016).

53 "two brothers or eight cousins": John Maynard Smith, "Survival through Suicide," *New Scientist*, August 28, 1975, 496.

53 There are other forms: Robert L. Trivers, "The Evolution of Reciprocal Altruism," *Quarterly Review of Biology* 46, no. 1 (1971): 36–57.

55 "If an individual's present": Denys deCatanzaro, "Human Suicide: A Biological Perspective," *Behavioral and Brain Sciences* 3, no. 265 (1980): 271.

56 "Constant anhedonia is a hallmark": Paul J. Watson and Paul W. Andrews, "Toward a Revised Evolutionary Adaptationist Analysis of Depression: The Social Navigation Hypothesis," *Journal of Affective Disorders* 72, no. 1 (2002): 5.

58 "Depression should abate": Ibid., 4.

58 "Even when a therapist": Ibid., 11.

59 "As an honest signal": Ibid., 7.

60 "social bargaining hypothesis": Kristen L. Syme, Zachary H. Garfield, and Edward H. Hagen, "Testing the Bargaining vs. Inclusive Fitness Models of Suicidal Behavior against the Ethnographic Record," *Evolution & Human Behavior* 37, no. 3 (2016): 179–92.

60 "Rather than altering": Ibid., 191.

61 baby really screaming: Silvia M. Bell and Mary D. Salter Ainsworth, "Infant Crying and Maternal Responsiveness," *Child Development* 43, no. 4 (1972): 1171–90.

61 "The prototypical suicidal state": Edwin S. Shneidman, *The Suicidal Mind* (Oxford: Oxford University Press, 1996), 133.

62 woman who'd been treated: Kristian Petrov, "The Art of Dying as an Art of Living: Historical Contemplations on the Paradoxes of Suicide and the Possibilities of Reflexive Suicide Prevention," *Journal of Medical Humanities* 34, no. 3 (2013): 347–68.

62 The most common assumption: Valerie J. Callanan and Mark S. Davis, "Gender Differences in Suicide Methods," *Social Psychiatry and Psychiatric Epidemiology* 47, no. 6 (2012): 857–69.

63 "For our custom up here": Knud Rasmussen, *The Netsilik Eskimos: Social*

Life and Spiritual Culture (Report of the Fifth Thule Expedition) (Copenhagen: Gyldendalske Boghandel, Nordisk Forlag, 1931), 143–44.

64 Japanese Kamikaze fighter pilots: John Orbell and Tomonori Morikawa, "An Evolutionary Account of Suicide Attacks: The Kamikaze Case," *Political Psychology* 32, no. 2 (2011): 297–322.

64 "Adaptive sense would appear": Ibid., 318.

65 kids still live at home: Katie Driver and Riadh T. Abed, "Does Having Offspring Reduce the Risk of Suicide in Women?" *International Journal of Psychiatry in Clinical Practice* 8, no. 1 (2004): 25–29.

65 "Whatever may be the positive": Knud J. Helsing and Mary Monk, "Dog and Cat Ownership among Suicides and Matched Controls," *American Journal of Public Health* 75, no. 10 (1985): 1224.

67 "an exemplar of psychopathology": Thomas E. Joiner et al., "Suicide as a Derangement of the Self-Sacrificial Aspect of Eusociality," *Psychological Review* 123, no. 3 (2016): 235.

67 "unsanctioned and frequently brutal": Ibid., 8.

67 For a fascinating historical: Ian Marsh, "The Uses of History in the Unmaking of Modern Suicide," *Journal of Social History* 46, no. 3 (2013): 744–56.

68 When given the choice: Katherine D. Kinzler, Emmanuel Dupoux, and Elizabeth S. Spelke, "The Native Language of Social Cognition," *Proceedings of the National Academy of Sciences* 104, no. 3 (2007): 12577–80.

68 "social distance scale": Richard A. Kalish, "Social Distance and the Dying," *Community Mental Health Journal* 2, no. 2 (1966): 152–55.

69 the trends were identical: David Lester, "The Stigmas against Dying and Suicidal Patients: A Replication of Richard Kalish's Study Twenty-Five Years Later," *OMEGA—Journal of Death and Dying* 26, no. 1 (1993): 71–75.

69 "Will the stigma be attached": Mark I. Solomon, "The Bereaved and the Stigma of Suicide," *OMEGA—Journal of Death and Dying* 13, no. 4 (1982): 385.

69 remarkably ant-like: Martin A. Nowak, Corina E. Tarnita, and Edward O. Wilson, "The Evolution of Eusociality," *Nature* 466, no. 7310 (2010): 1057–62.

70 "eusocial species": Thomas E. Joiner et al., "Suicide as a Derangement of the Self-Sacrificial Aspect of Eusociality," *Psychological Review* 123, no. 3 (2016): 237.

70 It abandons the colony: Robert Poulin, "Altered Behaviour in Parasitized Bumblebees: Parasite Manipulation or Adaptive Suicide?" *Animal Behaviour* 44, no. 1 (1992): 174–76.

70 "The adoption of": Ibid., 175.

71 "This misperception is itself": Thomas E. Joiner et al., "Suicide as a

Derangement of the Self-Sacrificial Aspect of Eusociality," *Psychological Review* 123, no. 3 (2016): 244.

73 some small island communities: Bronisław Malinowski, *Crime and Custom in Savage Society* (London: Routledge and Kegan Paul, 1926).

73 Under such conditions: Erving Goffman, *Stigma: Notes on the Management of Spoiled Identity* (1963; New York: Simon and Schuster, 2009).

75 16 percent of American: Kimberly J. Mitchell et al., "Exposure to Websites That Encourage Self-Harm and Suicide: Prevalence Rates and Association with Actual Thoughts of Self-Harm and Thoughts of Suicide in the United States," *Journal of Adolescence* 37, no. 8 (2014): 1335–44.

78 "Listen to me": Jean-Jacques Rousseau, *Julie, or the New Heloise: Letters of Two Lovers Who Live in a Small Town at the Foot of the Alps*, vol. 6, ed. Philip Steward and Jean Vaché (1761; Hanover, NH: University Press of New England, 2010), 323.

79 plummet during wartime: Stephen J. Rojcewicz, "War and Suicide," *Suicide and Life-Threatening Behavior* 1, no. 1 (1971): 46–54.

79 "To be rooted": Simone Weil, *The Need for Roots: Prelude to a Declaration of Duties towards Mankind* (1949; Oxford: Routledge Classics, 2001), 43.

79 rates in Nazi concentration camps: David Lester, *Suicide and the Holocaust* (Hauppauge, NY: Nova Science Publishers, 2005).

79 calls into question the low-suicide axiom: Francisco López-Muñoz and Esther Cuerda-Galindo, "Suicide in Inmates in Nazi and Soviet Concentration Camps: Historical Overview and Critique," *Frontiers in Psychiatry* 7, no. 88 (2016): 1–6.

80 "A domestic animal lives": Richard Dawkins, "Domesticity, Senescence, and Suicide," *Behavioral and Brain Sciences* 3, no. 2 (1980): 274.

80 Looking at twin studies: David A. Brent and J. John Mann, "Family Genetic Studies, Suicide, and Suicidal Behavior," *American Journal of Medical Genetics* 133, no. 1 (2005): 13–24.

81 "When did a conscious": Kay Redfield Jamison, *Night Falls Fast: Understanding Suicide* (New York: Vintage, 2000), 12.

82 "It is the opinion": S. R. Steinmetz, "Suicide among Primitive Peoples," *American Anthropologist* 7, no. 1 (1894): 53–60.

82 "It seems probable": Ibid., 60.

83 Instances of suicide emerged: Kristen L. Syme, Zachary H. Garfield, and Edward H. Hagen, "Testing the Bargaining vs. Inclusive Fitness Models of Suicidal Behavior against the Ethnographic Record," *Evolution & Human Behavior* 37, no. 3 (2016): 179–92.

84 Jessica Choi yuk-Chun: Simon Parry, "Taking the Easy Way Out?" *South China Morning Post*, January 9, 2005, http://www.scmp.com/article/484827/taking-easy-way-out.

84 Within a few short years: Kathy P. M. Chan et al., "Charcoal-Burning Suicide in Post-Transition Hong Kong," *British Journal of Psychiatry* 186, no. 1 (2005): 67–73.

84 "It is a suffocating": Parry, "Taking the Easy Way Out?"

85 "Any general theory of suicide": Denys deCatanzaro, "Human Suicide: A Biological Perspective," *Behavioral and Brain Sciences* 3, no. 265 (1980): 272.

85 "Consider Boston lead singer": " 'I Am a Lonely Soul,' Delp's Suicide Note Says," MSNBC, March 15, 2007, http://www.today.com/id/17613903#.WcrZlGUZ_do.

CHAPTER FOUR: HACKING THE SUICIDAL MIND

86 "There are moments": T. S. Eliot, "Literature and the Modern World," *American Prefaces* 1 (1935): 20.

87 I was suffering from: Dorothy Tennov, *Love and Limerence: The Experience of Being in Love* (1979; Lanham, MD: Scarborough House, 1999).

87 tend to occur: Craig A. Hill, Judith E. Blakemore, and Patrick Drumm, "Mutual and Unrequited Love in Adolescence and Young Adulthood," *Personal Relationships* 4, no. 1 (1997): 15–23.

89 "torture and extremity": Robert Burton, *Anatomy of Melancholy* (1621; New York: New York Review Books, 2001), 431.

89 happened upon an old article: Roy F. Baumeister, "Suicide as Escape from Self," *Psychological Review* 97, no. 1 (1990): 90–113.

93 better-than-average lives: Michael Argyle, *The Psychology of Happiness* (London: Methuen, 1987).

93 former professional rugby player: "Daniel Vickerman, Former Australia Lock, Dies 37," *Telegraph*, February 19, 2007, http://www.telegraph.co.uk/rugby-union/2017/02/19/dan-vickerman-former-australia-lock-dies-aged-37/.

93 the boy had stumbled: "South Carolina Father Commits Suicide after His 2-Year-Old Finds Loaded Gun, Fatally Shoots Himself," *KTLA 5 News*, September 7, 2017, http://ktla.com/2017/09/07/father-commits-suicide-after-2-year-old-finds-loaded-gun-shoots-himself/.

94 noticeably creep up: Herbert Hendin, *Suicide in America* (New York: Norton, 1982).

95 married to being single: Marianne Wyder, Patrick Ward, and Diego De Leo, "Separation as a Suicide Risk Factor," *Journal of Affective Disorders* 116, no. 3 (2009): 208–13.

95 most suicides in jails: Simon A. Backett, "Suicide in Scottish Prisons," *British Journal of Psychiatry* 151, no. 2 (1987): 218–21.

97 Most of us see ourselves: Robert Trivers, *The Folly of Fools: The Logic of Deceit and Self-Deception in Human Life* (New York: Basic Books, 2011).

97 "It is perhaps of little": Nicholas Epley and Erin Whitchurch, "Mirror, Mirror on the Wall: Enhancement in Self-Recognition," *Personality and Social Psychology Bulletin* 34, no. 9 (2008): 1169.

102 "best route to understanding": Edwin S. Shneidman, *The Suicidal Mind* (Oxford: Oxford University Press, 1996), 6.

102 "his file contained a memo": Susanne Langer, Jonathan Scourfield, and Ben Fincham, "Documenting the Quick and the Dead: A Study of Suicide Case Files in a Coroner's Office," *Sociological Review* 56, no. 2 (2008): 304.

103 Psycholinguists believe this is indicative: John Pestian et al., "Suicide Note Classification Using Natural Language Processing: A Content Analysis," *Biomedical Informatics Insights* 2010, no. 3 (2010): 19–28.

103 Perhaps the best study: Valerie J. Callanan and Mark S. Davis, "A Comparison of Suicide Note Writers with Suicides Who Did Not Leave Notes," *Suicide and Life-Threatening Behavior* 39, no. 5 (2009): 558–68.

104 "I have the feeling": Robert Wennersten, "Paying for Horses" (Interview with Charles Bukowski), *London Magazine*, December 1974, http://www.enotes.com/topics/charles-bukowski/critical-essays/bukowski-charles-vol-108#critical-essays-bukowski-charles-vol-108-criticism-robert-wennersten-interview-date-december-1974.

104 "The important thing was": Al Alvarez, *The Savage God: A Study of Suicide* (1971; London: Bloomsbury, 2002), 295.

105 "The so-called 'psychotically depressed'": David Foster Wallace, *Infinite Jest* (New York: Back Bay Books, 1997), 696.

106 a fox that has been: Sabar Rustomjee, "The Solitude and Agony of Unbearable Shame," *Group Analysis* 42, no. 2 (2009): 143–55.

106 "I could not handle": Edwin S. Shneidman, *The Suicidal Mind* (Oxford: Oxford University Press, 1996), 74.

107 Psychodynamic theorists: Herbert Hendin, "The Psychodynamics of Suicide," *Journal of Nervous and Mental Disease* 136, no. 3 (1963): 236–44.

108 "When I think about it": Chikako Ozawa-De Silva, "Shared Death: Self, Sociality and Internet Group Suicide in Japan," *Transcultural Psychiatry* 47, no. 3 (2010): 397.

108 "Should I kill myself": Giovanni Gaetani, "The Noble Art of Misquoting Camus: From Its Origins to the Internet Era," *Journal of Camus Studies* 1 (2015): 45.

109 With cognitive deconstruction: Robin R. Vallacher and Daniel M. Wegner, "What Do People Think They're Doing? Action Identification and Human Behavior," *Psychological Review* 94, no. 1 (1987): 3–15.

109 "a narrow, unemotional focus": Roy F. Baumeister, "Suicide as Escape from Self," *Psychological Review* 97, no. 1 (1990): 99–100.

109 "resemble acutely bored people": Ibid., 100.

110 state of flat affect: James W. Pennebaker, "Stream of Consciousness and Stress: Levels of Thinking," in *The Direction of Thought: Limits of Awareness, Intention, and Control*, ed. James S. Uleman and John A. Bargh (New York: Guilford, 1989), 327–50.

110 "[It] comes to resemble": William Styron, *Darkness Visible* (1990; London: Vintage Books, 2004), 49–50.

110 One old study: Louis A. Gottschalk and Goldine C. Gleser, "An Analysis of the Verbal Content of Suicide Notes," *Psychology and Psychotherapy: Theory, Research and Practice* 33, no. 3 (1960): 195–204.

110 "The typical purpose": Natalie J. Jones and Craig Bennell, "The Development and Validation of Statistical Prediction Rules for Discriminating between Genuine and Simulated Suicide Notes," *Archives of Suicide Research* 11, no. 2 (2007): 219–33.

111 "When preparing for suicide": Roy F. Baumeister, "Suicide as Escape from Self," *Psychological Review* 97, no. 1 (1990): 108.

111 losing a patient to suicide: Allison J. Darden and Philip A. Rutter, "Psychologists' Experiences of Grief after Client Suicide: A Qualitative Study," *OMEGA—Journal of Death and Dying* 63, no. 4 (2011): 317–42.

111 "The patient showed signs": Ibid., 325.

111 "with ironic tranquility": Émile Durkheim, *Suicide: A Study in Sociology*, trans. John A. Spaulding (1897; New York: Free Press, 1997), 283.

112 21,334: Kerry Shaw, "10 Essential Facts about Guns and Suicide," *The Trace*, September 6, 2016, https://www.thetrace.org/2016/09/10-facts-guns-suicide-prevention-month/.

112 "the single most dangerous": Edwin S. Shneidman, *The Suicidal Mind* (Oxford: Oxford University Press, 1996), 59.

113 "acquired capability for suicide": Kimberly A. Van Orden et al., "The Interpersonal Theory of Suicide," *Psychological Review* 117, no. 2 (2010): 575–600.

113 "Suicide is like diving": Al Alvarez, *The Savage God: A Study of Suicide* (1971; London: Bloomsbury, 2002), 108.

113 "To kill themselves": Kay Redfield Jamison, *Night Falls Fast: Understanding Suicide* (New York: Vintage, 2000), 133.

114 suicides in the U.S. military: Kimberly A. Van Orden et al., "Suicidal Desire and the Capability for Suicide: Tests of the Interpersonal-Psychological Theory of Suicidal Behavior among Adults," *Journal of Consulting and Clinical Psychology* 76, no. 1 (2008): 72–83.

CHAPTER FIVE: THE THINGS SHE TOLD LORRAINE

116 "It appears to me": Madame de Staël, *Reflections on Suicide* (1813), in *The Constitution of Man, Considered in Relation to Eternal Objects*, Alexandrian

Edition, ed. George Combe (Columbus, OH: J & H Miller, n.d.), 99.

123 1970s crooner Barry Manilow: Thomas Gilovich, Victoria H. Medvec, and Kenneth Savitsky, "The Spotlight Effect in Social Judgment: An Egocentric Bias in Estimates of the Salience of One's Own Actions and Appearance," *Journal of Personality and Social Psychology* 78, no. 2 (2000): 211–22.

129 ability to envision the future: Charles Neuringer and Robert M. Harris, "The Perception of the Passage of Time among Death-Involved Hospital Patients," *Suicide and Life-Threatening Behavior* 4, no. 4 (1974): 240–54.

140 "Who among us has never": Kai Epstude and Neal J. Roese, "The Functional Theory of Counterfactual Thinking," *Personality and Social Psychology Review* 12, no. 2 (2008): 168.

140 "Counterfactual thoughts are mental": Ibid., 168.

CHAPTER SIX: TO LOG OFF THIS MORTAL COIL

145 "All the inconveniences": Michel de Montaigne, *A Custom of the Isle of Cea* (1574), ed. W. C. Hazlitt, trans. Charles Cotton (1686; Kensington, 1877). Available online at www.gutenberg.org, the Gutenberg Project, text #3600.

146 "There are many reasons": Nic Sheff, "13 Reasons Why Writer: Why We Didn't Shy Away from Hannah's Suicide," *Vanity Fair*, April 19, 2017, https://www.vanityfair.com/hollywood/2017/04/13-reasons-why-suicide-controversy-nic-sheff-writer.

147 About thirty years before: Armin Schmidtke and Heinz Häfner, "The Werther Effect after Television Films: New Evidence for an Old Hypothesis," *Psychological Medicine* 18, no. 3 (1988): 665–76.

147 After Goethe first published: Johann Goethe, *The Sorrows of Young Werther*, vol. 10 (1774; London: Penguin, 2006).

148 single-car road fatalities: David P. Phillips, "Suicide, Motor Vehicle Fatalities, and the Mass Media: Evidence toward a Theory of Suggestion," *American Journal of Sociology* 84, no 5 (1979): 1150–74.

148 In July 2017: John W. Ayers et al., "Internet Searches for Suicide Following the Release of *13 Reasons Why*," *JAMA Internal Medicine*. July 31, 2017, doi:10.1001/jamainternmed.2017.3333.

149 "In relative terms": Sarah Knapton, "Netflix Series '13 Reasons Why' Should Be Withdrawn after Triggering Spike in 'How to Commit Suicide' Searches," *Telegraph*, July 31, 2017, http://www.telegraph.co.uk/science/2017/07/31/netflix-series-13-reasons-should-withdrawn-triggering-spike/.

149 shown a billboard: Bonnie Klimes-Dougan et al., "Suicide Prevention with Adolescents: Considering Potential Benefits and Untoward Effects of Public Service Announcements," *Crisis* 30, no. 3 (2009): 128–35.

149 "The question of how": Thomas Niederkrotenthaler et al., "Increasing

Help-Seeking and Referrals for Individuals at Risk for Suicide by Decreasing Stigma: The Role of Mass Media," *American Journal of Preventive Medicine* 47, no. 3 (2014): S239.

150 a sort of meta-review: Steven Stack, "Media Coverage as a Risk Factor in Suicide," *Injury Prevention* 8, no. 4 (2002): iv30–iv32.

150 "If Marilyn Monroe": Ibid., iv31.

151 helpful list of do's and don'ts: "Recommendations for Reporting on Suicide," http://reportingonsuicide.org/wp-content/themes/ros2015/assets/images/Recommendations-eng.pdf.

154 search terms used: Paul W. C. Wong et al., "Accessing Suicide-Related Information on the Internet: A Retrospective Observational Study of Search Behavior," *Journal of Medical Internet Research* 15, no. 1 (2013): e3.

155 "The expressions of violence": Michael Westerlund, "The Production of Pro-Suicide Content on the Internet: A Counter-Discourse Activity," *New Media & Society* 14, no. 5 (2012): 773.

156 The Japanese media: Yoshihiro Nabeshima et al., "Analysis of Japanese Articles about Suicides Involving Charcoal Burning or Hydrogen Sulfide Gas," *International Journal of Environmental Research and Public Health* 13, no. 10 (2016): 1–12.

159 The young woman who'd: Hidenori Tomita, "*Keitai* and the Intimate Stranger," in *Personal, Portable, Pedestrian: Mobile Phones in Japanese Life*, ed. Mizuko Ito, Misa Matsuda, and Daisuke Okabe (Cambridge, MA: MIT Press, 2005), 183–201.

159 "Persons who wish to": Chikako Ozawa-de Silva, "Too Lonely to Die Alone: Internet Suicide Pacts and Existential Suffering in Japan," *Culture, Medicine, and Psychiatry* 32, no. 4 (2008): 537.

159 In South Korea: David D. Luxton, Jennifer D. June, and Jonathan M. Fairall, "Social Media and Suicide: A Public Health Perspective," *American Journal of Public Health* 102, no. 2S2 (2012): S195–S200.

159 England's first publicized case: Ian Cobain, "Clampdown on Chatrooms after Two Strangers Die in First Internet Death Pact," *Guardian*, October 11, 2005, https://www.theguardian.com/uk/2005/oct/11/socialcare.technology.

160 Prior to the internet: Patricia A. Santy, "Observations on Double Suicide: Review of the Literature and Two Case Reports," *American Journal of Psychotherapy* 36, no. 1 (1982): 23–31.

160 intent remains somewhat unclear: "Net Grief for Online 'Suicide,'" *BBC News*, February 4, 2013, http://news.bbc.co.uk/2/hi/technology/2724819.stm.

161 Here's a small sample: Paul Harper and Gemma Mullin, "Death on Camera: How Facebook Live Murder and Suicide Videos Are Spreading

Online and What to Do if You Spot Inappropriate Content," *The Sun*, July 10, 2017, https://www.thesun.co.uk/news/3426352/facebook-live-clips-murder-suicide-shootings-report/.

161 "not just about dying": Michael Westerlund, Gergö Hadlaczky, and Danuta Wasserman, "Case Study of Posts Before and After a Suicide on a Swedish Internet Forum," *British Journal of Psychiatry* 207, no. 6 (2015): 480.

162 a unique case study: Ibid.

163 on the autism spectrum: Sarah Cassidy et al., "Suicidal Ideation and Suicide Plans or Attempts in Adults with Asperger's Syndrome Attending a Specialist Diagnostic Clinic: A Clinical Cohort Study," *Lancet Psychiatry* 1, no. 2 (2014): 142–47.

163 "could suggest a different": Ibid., 146.

164 "The act of suicide": Joost A. M. Meerloo, "Suicide, Menticide, and Psychic Homicide," *AMA Archives of Neurology & Psychiatry* 81, no. 3 (1959): 112.

164 prosecution of Michelle Carter: Katherine Q. Seelye, "Michelle Carter Gets 15-Month Jail Term in Texting Suicide Case," *New York Times*, August 3, 2017, https://www.nytimes.com/2017/08/03/us/texting-suicide-sentence.html?mcubz=3.

166 "These pages typically included": Lindsay Robertson et al., "An Adolescent Suicide Cluster and the Possible Role of Electronic Communication Technology," *Crisis* 33 (2012): 241.

166 "Sources reported that young": Ibid., 242.

166 And finally, cyberbullying: Mitch Van Geel, Paul Vedder, and Jenny Tanilon, "Relationship between Peer Victimization, Cyberbullying, and Suicide in Children and Adolescents: A Meta-Analysis," *JAMA Pediatrics* 168, no. 5 (2014): 435–42.

167 One recent meta-analysis: Ibid.

167 has become so problematic: Centers for Disease Control and Prevention, "Youth Violence: Technology and Youth—Protecting Your Child from Aggression" (2014), http://www.cdc.gov/violenceprevention/pdf/ea-tipsheet-a.pdf.

168 "mother of all fears": Philippe Rochat, *Others in Mind: Social Origins of Self-Consciousness* (Cambridge: Cambridge University Press, 2009), 21.

168 illustrating our extraordinary sensitivity: Chris H. J. Hartgerink et al., "The Ordinal Effects of Ostracism: A Meta-Analysis of 120 Cyberball Studies," *PLOS ONE* 10, no. 5 (2015): e0127002.

169 "How can an otherwise": Matthieu J. Guitton, "The Importance of Studying the Dark Side of Social Networks," *Computers in Human Behavior* 31 (2014): 355.

169 The positive influence of media: Thomas Niederkrotenthaler et al., "Role of Media Reports in Completed and Prevented Suicide: Werther v. Papag-

eno Effects," *British Journal of Psychiatry* 197, no. 3 (2010): 234–43.

170 "Using AI to identify": Prathamesh Mulye, "Artificial Intelligence: Future Perfect, Future Tense," *Deccan Chronicle*, August 27, 2017, http://www.deccanchronicle.com/decaf/270817/artificial-intelligence-future-perfect-future-tense.html.

172 "the only thing separating": J. L. Long, "Logotherapeutic Transcendental Crisis Intervention," *International Forum for Logotherapy* 20, no. 4 (1997): 107.

CHAPTER SEVEN: WHAT DOESN'T DIE

175 "I'd rather there wasn't": Bruce Lambert, William Golding Is Dead at 81; The Author of 'Lord of the Flies,'" *New York Times*, June 20, 1993, http://www.nytimes.com/learning/general/onthisday/bday/0919.html.

177 "You might as well live": Dorothy Parker, "Résumé," in *The Portable Dorothy Parker* (New York: Penguin, 1976), 99.

180 "[I] regard the brain": Ian Sample, "Stephen Hawking: 'There Is No Heaven: It's a Fairy Story,'" *Guardian*, May 15, 2011, https://www.theguardian.com/science/2011/may/15/stephen-hawking-interview-there-is-no-heaven.

180 "hard problem of consciousness": David Chalmers, e-mail message to author, June 10, 2017.

181 "You're nothing but": Francis Crick, *The Astonishing Hypothesis: The Scientific Search for the Soul* (New York: Scribner, 1995), 3.

182 I conducted a study: Jesse M. Bering, "Intuitive Conceptions of Dead Agents' Minds: The Natural Foundations of Afterlife Beliefs as Phenomenological Boundary," *Journal of Cognition and Culture* 2, no. 4 (2002): 263–308.

183 similar study with young children: Jesse M. Bering and David F. Bjorklund, "The Natural Emergence of Reasoning about the Afterlife as a Developmental Regularity," *Developmental Psychology* 40, no. 2 (2004): 217–33.

184 "Everywhere in the snarl": Kay Redfield Jamison, *Night Falls Fast: Understanding Suicide* (New York: Vintage, 2000), 182.

184 evolutionary biologists: Daniel J. Povinelli and John G. H. Cant, "Arboreal Clambering and the Evolution of Self-Conception," *Quarterly Review of Biology* 70, no. 4 (1995): 393–421.

186 "I stayed a long time": Andre Gide, *The Counterfeiters: A Novel* (1927; New York: Vintage Books, 1973), 249.

186 "Hostility Directed Outwards Type": Jacob Tuckman, Robert J. Kleiner, and Martha Lavell, "Emotional Content of Suicide Notes," *American Journal of Psychiatry* 116, no. 1 (1959): 59–63.

187 "Our life has no end": Ludwig Wittgenstein, *Tractatus Logico-Philosophicus*,

trans. David F. Pears and Bernard F. McGuinness (London: Routledge and Kegan Paul, 1961), 6:4311.

187 "Our own death": Sigmund Freud, "Thoughts for the Times on War and Death," in *Collected Works of C. G. Jung*, ed. Herbert Read, Michael Fordham, and Gerhard Adler, vol. 4, *Freud and Psychoanalysis* (1913; London: Hogarth, 1953), 304–5.

187 "try to fill your consciousness": Miguel de Unamuno, *Tragic Sense of Life*, trans. J. E. Crawford Flitch (1912; Charleston, SC: BiblioBazaar, 2007), 71.

188 "imagine my own non-existence": Shaun Nichols, "Imagination and Immortality: Thinking of Me," *Synthese* 159 (2007): 220.

188 Consider just one of: Thomas Fields-Meyer, "A Fatal Passion," *People*, November 27, 1995, http://people.com/archive/a-fatal-passion-vol-44-no-22/.

189 "A yearning for death": Plutarch, *Moralia*: "Bravery of Women," XI, trans. Frank C. Babbitt (Cambridge, MA: Harvard University Press, 1949), 509.

189 "There is nothing, once": Thomas Lynch, *The Undertaking: Life Studies from the Dismal Trade* (1997; New York: Norton, 2009), 7.

190 That was the poet: Dante Alighieri, *The Inferno*, trans. John Ciardi (New York: Signet, 1954).

191 "is so monstrous": John Sym, *Life's Preservative against Self-Killing*, ed. Michael MacDonald (1637; London: Routledge, 1988), 53.

191 "grave psychological disturbances": Catholic Church, "Suicide," in *Catechism of the Catholic Church*, 2nd ed. (Vatican: Libreria Editrice Vaticana, 2012).

191 "We are stewards": Ibid.

192 Adherents of this movement: Horst J. Koch, "Suicides and Suicide Ideation in the Bible: An Empirical Survey," *Acta Psychiatrica Scandinavica* 112, no. 3 (2005): 167–72.

192 we see purpose everywhere: Deborah Kelemen, "Are Children 'Intuitive Theists'? Reasoning about Purpose and Design in Nature," *Psychological Science* 15, no. 5 (2004): 295–301.

192 "I condemn that nature": Fyodor Dostoyevsky, *The Diary of a Writer*, vol. 3 (1877; London: Charles Scribner's Sons, 1949).

193 Only 1.6 percent: James C. Harris, "Ophelia," *Archives of General Psychiatry* 64, no. 10 (2007): 1114.

193 "The belief that suicide": Ibid.

193 "Halakhic law stipulated": Yari Gvion, Yossi Levi-Belz, and Alan Apter, "Suicide in Israel—An Update," *Crisis* 35, no. 3 (2014): 142.

193 "the soul has nowhere": Robin E. Gearing and Dana Lizardi, "Religion and Suicide," *Journal of Religion and Health* 48, no. 3 (2009): 337.

194 the Quran does not include: Franz Rosenthal, "On Suicide in Islam," *Journal of the American Oriental Society* 66 (1946): 239–59.

194 "He who kills himself": Ibid., 244.

195 "We believe that no": Interview with Osama bin Laden, CNN, March 1997, http://news.findlaw.com/hdocs/docs/binladen/binladenintvw-cnn.pdf.

195 "Each time we sleep": Ibn Warraq, "Virgins? What Virgins?" *Guardian*, January 12, 2002, https://www.theguardian.com/books/2002/jan/12/books.guardianreview5.

195 Although Hindu scriptures: Lata Mani, *Contentious Traditions: The Debate on Sati in Colonial India* (Berkeley: University of California Press, 1998).

195 continuing into modern times: "Indian Women Still Commit Ritual Suicides," *RT*, September 9, 2009, https://www.rt.com/news/india-ritual-suicide-sati/.

196 "A wife who dies": Jagdish L. Shastri and G. P. Bhatt, *Garuda Purana Pt. 1, Ancient Indian Tradition and Mythology*, vol. 12 (Motilal Banarsidass Publishers Pvt. Ltd., 2008), 1.107.29.

196 And speaking of gender: Jie Zhang and Huilan Xu, "The Effects of Religion, Superstition, and Perceived Gender Inequality on the Degree of Suicide Intent: A Study of Serious Attempters in China," *OMEGA—Journal of Death and Dying* 55, no. 3 (2007): 185–97.

196 "To some Chinese individuals": Ibid., 189–90.

197 later forms of canonical: Dana Lizardi and Robin E. Gearing, "Religion and Suicide: Buddhism, Native American and African Religions, Atheism, and Agnosticism," *Journal of Religion and Health* 49, no. 3 (2010): 377–84.

197 In the late 1980s: Zhou Juhua, "A Probe into the Mentality of Sixty-Five Rural Young Women Giving Birth to Baby Girls," *Chinese Sociology & Anthropology* 20, no. 3 (1988): 93–100.

198 Ever since Durkheim: Émile Durkheim, *Suicide: A Study in Sociology*, trans. John A. Spaulding (1897; New York: Free Press, 1997).

198 "network theory": Nichole C. Rushing et al., "The Relationship of Religious Involvement Indicators and Social Support to Current and Past Suicidality among Depressed Older Adults," *Aging & Mental Health* 17, no. 3 (2013): 366–74.

198 "Religious participation and network contacts": Ibid., 366.

198 "This powerful narrative": William K. Kay and Leslie J. Francis, "Suicidal Ideation among Young People in the UK: Churchgoing as an Inhibitory Influence?" *Mental Health, Religion & Culture* 9, no. 2 (2006): 133.

199 Several studies have shown: Erminia Colucci and Graham Martin, "Religion and Spirituality along the Suicidal Path," *Suicide and Life-Threatening Behavior* 38, no. 2 (2008): 229–44.

199 Religion or no: Joshua Rottman, Deborah Kelemen, and Liane Young, "Tainting the Soul: Purity Concerns Predict Moral Judgments of Suicide," *Cognition* 130, no. 2 (2014): 217–26.

199 "Suicide represents an unusual": Ibid., 217.

201 Related to Rottman's findings: O. Henry, *The Furnished Room*, ed. Siegfried Schmitz (1904; London: Royal Blind Society, 1980).

201 Real life is even: Stuart C. Edmistron, "Secrets Worth Keeping: Toward a Principled Basis for Stigmatized Property Disclosure Statutes," *UCLA Law Review* 58 (2010): 281–320.

201 a couple in Auckland: "Agents Win Appeal over House Suicide Secret," *New Zealand Herald*, November 18, 2014, http://www.nzherald.co.nz/business/news/article.cfm?c_id=3&objectid=11360526.

202 participants abjectly refused: Paul Rozin and Edward B. Royzman, "Negativity Bias, Negativity Dominance, and Contagion," *Personality and Social Psychology Review* 5, no. 4 (2001): 296–320.

203 the body is removed: Mensah Adinkrah, "Better Dead Than Dishonored: Masculinity and Male Suicidal Behavior in Contemporary Ghana," *Social Science & Medicine* 74, no. 4 (2012): 474–81.

203 In Japan, where dense: Ryo Seven, "Jiko Bukken: The Cheapest Apartments in Tokyo . . . but Only If You're Brave," *Tokyocheapo*, April 3, 2015, https://tokyocheapo.com/living/jiko-bukken-cheapest-apartments-in-tokyo/.

203 "I feel envious of": Mary Picone, "Suicide and the Afterlife: Popular Religion and the Standardisation of 'Culture' in Japan," *Culture, Medicine, and Psychiatry* 36, no. 2 (2012): 391–408.

206 In this study published: Jesse M. Bering, Emma R. Curtin, and Jonathan Jong, "Knowledge of Deaths in Hotel Rooms Diminishes Perceived Value and Elicits Guest Aversion," *OMEGA—Journal of Death and Dying* (2017): 0030222817709694.

CHAPTER EIGHT: GRAY MATTER

213 "Disgust with our own": Voltaire, "Cato: On Suicide, and the Abbe St. Cyrian's Book Legitimating Suicide," Online Library of Liberty, http://oll.libertyfund.org/titles/voltaire-the-works-of-voltaire-vol-iv-philosophical-dictionary-part-2?q=cato+on+suicide#.

213 "Do we ever say": Susan J. Beaton, Peter Forster, and Myfanwy Maple, "The Language of Suicide," *Psychologist* 25, no. 10 (2012): 731.

214 "My 'failed attempts,' ": Ibid.

214 "As health professionals": Ibid.

219 "This approach is limited": Matthew K. Nock et al., "Measuring the Sui-

cidal Mind: Implicit Cognition Predicts Suicidal Behavior," *Psychological Science* 21, no. 4 (2010): 511.

219 "The presence of an implicit": Ibid., 514–15.

222 "a wise man will": Seneca, *Ad Lucilium Epistulae Morales*, Letter 70, trans. Richard M. Gummere (New York: G. P. Putnam's Sons, 1920).

223 "If he disposes of": Immanuel Kant, *Lectures on Ethics*, trans. Louis Infield (New York: Harper & Row, 1978).

223 "slay his own best": *The Dialogues of Plato*, Laws IX, trans. Benjamin Jowett (New York: Random House, 1920).

224 "the life of a man": David Hume, "Of Suicide," manuscript in the National Library of Scotland with corrections in Hume's own hand, text provided by Tom L. Beauchamp; David Hume, "To John Home of Ninewells," in *The Letters of David Hume*, ed. John Y. T. Grieg (1757; Oxford: Clarendon Press, 1932).

224 I'm reminded of: Valerius Maximus, *Memorable Doings and Sayings*, Book II, 6th ed., trans. David R. Shackleton Bailey (Cambridge, UK: Loeb Classical Library, 2000).

225 "However great a man's": Graham Greene, *The Comedians* (1966; New York: Penguin Classics, 1991).

226 A few decades ago: Thomas Szasz, *The Myth of Mental Illness: Foundations of a Theory of Personal Conduct* (New York: Hoeber-Harper, 1961).

227 "a misleading slogan characteristic": Thomas Szasz, "The Case against Suicide Prevention," *American Psychologist* 41, no. 7 (1986): 808.

227 "Where there is no freedom": Ibid., 809.

227 Six months after Szasz: Melanie Hirsch, "Home on the Hot Seat," *Post-Standard*, February 19, 1992, http://www.syracuse.com/news/index.ssf/2012/09/dr_thomas_szasz_critics_discus.html.

227 "It would seem certain": Thomas Joiner, *Mindlessness: The Corruption of Mindfulness in a Culture of Narcissism* (Oxford: Oxford University Press, 2017), 88.

227 Would you identify as: Brian L. Mishara and David N. Weisstub, "Ethical and Legal Issues in Suicide Research," *International Journal of Law and Psychiatry* 28, no. 1 (2005): 23–41.

234 In a yearlong study: Garen J. Wintemute et al., "Mortality among Recent Purchases of Handguns," *New England Journal of Medicine* 341, no. 21 (1999): 1583–89.

234 "Hanging in most cases": Michael J. Kral, "Postcolonial Suicide among Inuit in Arctic Canada," *Culture, Medicine, and Psychiatry* 36, no. 2 (2012): 310.

234 Adding catalytic converters: Robert Evan Kendell, "Catalytic Converters and Prevention of Suicides," *Lancet* 352, no. 9139 (1998): 1525.

235 Bridge barriers erected: Annette L. Beautrais, "Effectiveness of Barriers at Suicide Jumping Sites: A Case Study," *Australian & New Zealand Journal of Psychiatry* 35, no. 5 (2001): 557–62.

235 "Many young suicide attempters": Urs Hepp et al., "Methods of Suicide Used by Children and Adolescents," *European Child & Adolescent Psychiatry* 21, no. 2 (2012): 72.

236 After a lonely Bay Area: Tad Friend, "Jumpers: The Fatal Grandeur of the Golden Gate Bridge," *New Yorker,* October 13, 2003, https://www.newyorker.com/magazine/2003/10/13/jumpers.

index